Current Research in Hydraulic Turbines VIII

Journal of Physics: Conference Series Volume 1042

Dhulikhel, Nepal
20 March 2018

ISBN: 978-1-5108-6493-1
ISSN: 1742-6588

Printed from e-media with permission by:

Curran Associates, Inc.
57 Morehouse Lane
Red Hook, NY 12571

CURRAN ASSOCIATES INC.
proceedings
.com

Some format issues inherent in the e-media version may also appear in this print version.

Printed by Curran Associates, Inc. (2018)

For permission requests, please contact the Institute of Physics at the address below.

Institute of Physics
Dirac House, Temple Back
Bristol BS1 6BE UK

Phone: 44 1 17 929 7481
Fax: 44 1 17 920 0979

techtracking@iop.org

Additional copies of this publication are available from:

Curran Associates, Inc.
57 Morehouse Lane
Red Hook, NY 12571 USA
Phone: 845-758-0400
Fax: 845-758-2633
Email: curran@proceedings.com
Web: www.proceedings.com

Table of contents

Volume 1042

Current Research in Hydraulic Turbines (CRHT) VIII

20 March 2018, Kathmandu University, Nepal

Accepted papers received: 25 May 2018
Published online: 12 June 2018

Preface

Papers

IOP Conf. Series: Journal of Physics: Conf. Series **1042** (2018) 011001 doi:10.1088/1742-6596/1042/1/011001

Preface

Current Research in Hydraulic Turbines (CRHT) is an international symposium dedicated to research activities related to hydraulic turbines. The event has been taking place in Kathmandu University since 2010 and since 2018, the publication of the short-listed papers from the event in IOP Journal of Physics: Conference Series has been provisioned. This year, the topic of the papers/presentations was divided into six thematic areas: i) Modeling and simulation in hydro turbines ii) Simulation and experimental techniques in hydropower iii) Sediment erosion and cavitation in hydraulic turbines iv) Renewable energy, devices, energy and efficiency measurement v) Pumps, FSI in pipe and net metering and vi) Reversible pump turbines and Hill chart of Francis turbine.

In this symposium, the technical session was represented by six Universities in three parallel sessions. In total, 24 papers were presented in the symposium, out of which, 10 papers were selected for publication in IOP. The selection was based on the evaluation and thorough review process by respective session chairs and the organizing committee. The event was organized by Turbine Testing Lab of Kathmandu University and supported by International Center for Hydropower (ICH).

The papers included in this Journal contain a wide range of research activities conducted in the sector of hydropower and hydraulic turbines. The problems of sediment erosion in the context of Indian Sub-continent have been discussed with the challenges of modeling and simulating the erosion in CFD. Theoretical approach for generating hill charts of Francis turbines using shroud-streamline modeling has been presented. Similarly, the problems of pressure pulsations in Francis turbines, experimental and numerical techniques of measurement and modeling of the power plants have been emphasized. Some papers also include design optimization using CFD for better efficiency and minimum erosion in turbines.

IOP Conf. Series: Journal of Physics: Conf. Series **1042** (2018) 011002 doi:10.1088/1742-6596/1042/1/011002

Peer review statement

All papers published in this volume of *Journal of Physics: Conference Series* have been peer reviewed through processes administered by the proceedings Editors. Reviews were conducted by expert referees to the professional and scientific standards expected of a proceedings journal published by IOP Publishing.

IOP Conf. Series: Journal of Physics: Conf. Series **1042** (2018) 012001 doi:10.1088/1742-6596/1042/1/012001

Performance Comparison of Optimized Designs of Francis Turbines Exposed to Sediment Erosion in various Operating Conditions

K P Shrestha[1*], S Chitrakar, B Thapa[1], O G Dahlhaug[2]

1 Department of Mechanical Engineering, Kathmandu University, Kavre, Nepal
[2]Department of Energy and Process Engineering, NTNU, Norway
*Corresponding author (kp@ku.edu.np)

Abstract. Erosion on hydro turbine mostly depends on impingement velocity, angle of impact, concentration, shape, size and distribution of erodent particle and substrate material. In the case of Francis turbines, the sediment particles tend to erode more in the off-designed conditions than at the best efficiency point. Previous studies focused on the optimized runner blade design to reduce erosion at the designed flow. However, the effect of the change in the design on other operating conditions was not studied. This paper demonstrates the performance of optimized Francis turbine exposed to sediment erosion in various operating conditions. Comparative study has been carryout among the five different shapes of runner, different set of guide vane and stay vane angles. The effect of erosion is studied in terms of average erosion density rate on optimized design Francis runner with Lagrangian particle tracking method in CFD analysis. The numerical sensitivity of the results are investigated by comparing two turbulence models. Numerical results are validated from the velocity measurements carried out in the actual turbine. Results show that runner blades are susceptible to more erosion at part load conditions compared to BEP, whereas for the case of guide vanes, more erosion occurs at full load conditions. Out of the five shapes compared, Shape 5 provides an optimum combination of efficiency and erosion on the studied operating conditions.

Keyword: Francis Turbine, Variable Speed, Pressure Pulsation, Hill Chart, Hydro Power

1. Introduction

Sediment problem is prominent in hydropower operating on run-off-river type projects. Erosion rate is higher during the monsoon period due to high concentration of sediment. Comparatively it is in the monsoon when the turbine components of the power plants suffer severe damage. Damage to the turbine components leads to the drop in efficiency. Eventually after a certain period of operation, it becomes necessary to shut down the plant for repair or for replacement of the damaged components [1] [2] [3] [4] [5] [6].

Karlen et. al., 2002 [7], proposed different types of hydro-erosive pattern for the evaluation of hydro turbines, such as metallic luster, fine-scaly erosion, scaly erosion, large sized scaly erosion, in depth scaly erosion and depth erosion. Brekke et.al, 2002 [8], classified the erosion phenomena into three different group as Turbulence erosion, acceleration erosion and secondary flow erosion. These categories of erosion are differentiated based on the location of impact and operating conditions of the turbine.

Jhimruk Hydropower plant located in Pyuthan district of Nepal is a run-off type power plant. It is an example of plants that suffer from severe problems due to sediments [2] [6]. It has three Francis Units, each of 4.2 MW. Figure 1 shows Francis turbine runner damaged by sediment erosion. An efficiency drop of 4% at BEP (Best Efficiency Point), within the period of 1 September 2003 and 11 November 2003 was studied during thermodynamic efficiency measurements of the plant, as presented in the figure [2] . Effect of sediments on turbine mainly depends upon flow parameter, sand particle characteristics and properties of material [1] [3] [9] [4] [10] [8]. Kaligandaki and Marsyangdi are some other examples of power plants in Nepal that suffer from severe damage due to sediments.

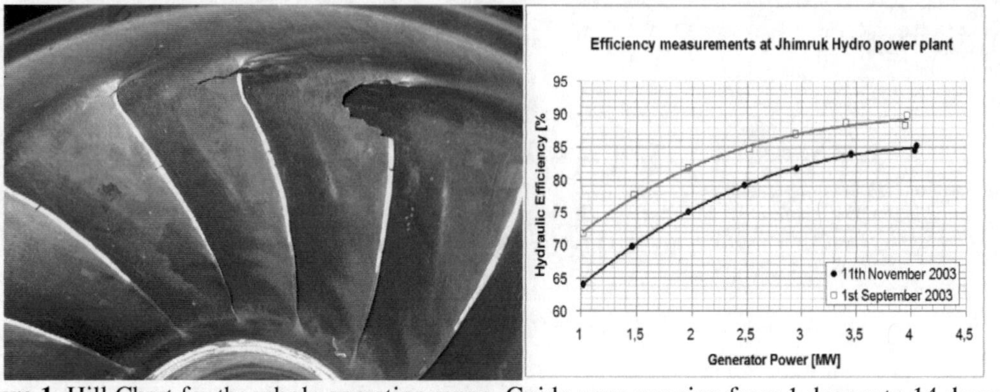

Figure 1. Hill Chart for the whole operation range. Guide vane opening from 1 degree to 14 degrees. NED varies from 0.08 to run away speed.

Neopane [3] showed that the erosion rate increases when turbine is operated in full load condition due to increase in turbulence and relative velocity at outlet. Several attempts have also been made to tackle the sediment problem using various methods of coating. The effectiveness of the coating to tackle sediment problem depends upon the spray parameters and powder properties [1] [12]. Coating done at Marsyangdi hydropower project and HVOF (High Velocity Oxygen Fuel) coating done at Kaligandaki 'A' reduced the erosion rate [6] [12], but in Jhimruk hydropower project , the coating was not of much effective as the coating did not even last for one season [6].

The conventional design of Francis turbine performs well in BEP in a clean water condition. However, erosion is more susceptible when operated in part and full load conditions [1] [3]. In the traditional design, most of hydraulic energy is converted at the beginning half of the blade [4] [13]. By changing the hydraulic design procedure, it has been shown is several studies that erosion can be reduced significantly [3] [4] [8] [10] [14] [15] [16] [17]. All of these studies gave priority to BEP, but there is a huge variation in the operating load and flow conditions. To maintain the constant speed of the turbine at different operations, the guide vane angle is adjusted. This adjustment changes the flow behavior, which eventually changes the nature and quantity of erosion in turbines. This study presents a numerical technique to investigate the performances of 5 designs of Francis turbine runner at several operating conditions.

2. Design methodology and numerical model

The basic design data of Jhimruk Power Plant, which is presented in Table 1, was taken as a reference case for developing new designs of Francis turbine [4] [10] [16] [17] [18] [19]. These design data were chosen for the initial design stage, and optimized design process was adopted to develop and evaluate the erosion impact on turbine.

Table 1. Basic Design Data for Jhimruk Power Plant [2] [16] [19]

S.N.	Parameters	Value	Unit
1	Net design head(H)	201.5	m
2	Net discharge per unit (Q)	2.35	m^3/s
3	Targeted Runner efficiency	96	%

CFD simulation was performed on the five different types of blades with different guide vane angles in wide range of operating conditions. These five set of runners were compared on the basis of pressure distribution, velocity gradient, velocities at leading and trailing edges. Lagrangian particle tracking method was applied to observe average erosion density rate on runner, guide vane and stay vane respectively.

CFD simulations were performed for 100 combinations of Guide Vane (GV) opening angles and the runner blade profiles. The estimation of erosion on stay vanes, guide vanes and runner blades were done on the basis on Lagrangian calculation of particle paths in a viscous flow. The original turbine contains 17 runner blades, 24 guide vanes and 24 stay vanes. However, considering the cyclic symmetricity of the region, a single blade passage, containing one stay vane, one guide vane and one runner blade was modeled for the numerical analysis in ANSYS CFX. This method reduced the computational time and the resources required. CFX is a commercial tool for computational studies that uses a finite-volume multi-block approach to solve the governing equations of fluid motion numerically on a computational grid [20]. In this study, k-epsilon and SST turbulence models based on RANS (Reynolds Averaged Navier-Stokes) equations were used in all the cases, as it offers a good combination of accuracy and robustness [4] [20]. The algebraic equations were solved iteratively with second order accurate approach. The simulations were steady state having viscous and incompressible flows.

The meshing was done in each of the fluid domains by using Turbo-grid. This is an in-built tool inside ANSYS, which creates high-quality hexahedral meshes to perform fluid dynamics analysis in rotating machinery. The factor ratio, or the rate of increase in the size of the mesh from boundary to the adjacent mesh for this analysis was chosen to be 1.15 so that the near wall refinements in the blades can be incorporated [4]. Increasing the value of the factor ratio makes it possible to meet the desired criteria of y+ value, but it will also have a convergence problem [4] [16] [17]. In this study, y+ value of ~30 was maintained around runner blade for all simulations. Although the recommended value of y+ for SST turbulence model is less than 1 according to ANSYS CFX guide, it was discussed in a previous study that the quality of the mesh degrades at such a low value of y+ in the case of Francis turbines [4].

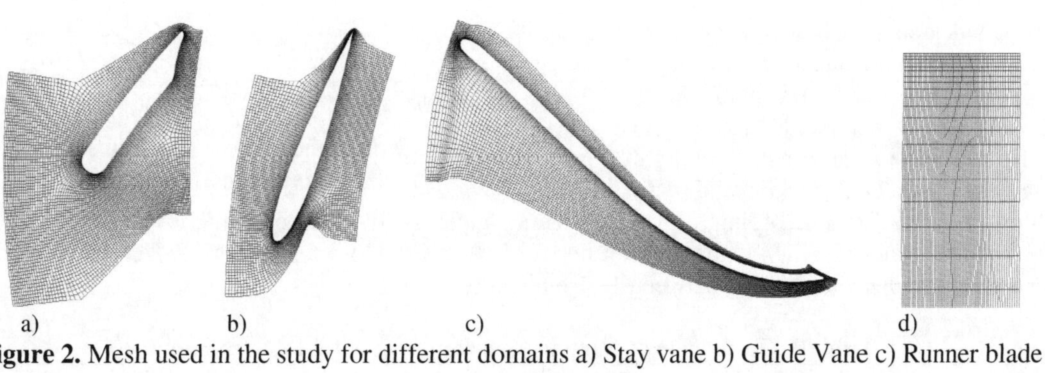

Figure 2. Mesh used in the study for different domains a) Stay vane b) Guide Vane c) Runner blade d) a portion of draft tube

Table 2. Number of mesh nodes used in all domains (typical case)

S.N.	Parameters	Value	Unit
1	Net design head(H)	201.5	m
2	Net discharge per unit (Q)	2.35	m^3/s
3	Targeted Runner efficiency	96	%

This study uses k-ε and SST turbulence models for all the analyses. k-ε model has been very successful in a large variety of flow regime in which has high Reynold Number without adverse pressure gradient [23]. SST model is useful for conditions where flow separation under adverse pressure gradient and gives more accuracy for the boundary layer simulations.

In k-ε model, Turbulent kinetic energy is defined as the variance of fluctuating velocity and turbulence eddy dissipation is calculate by rate at which velocity fluctuation dissipate. The value of k and ε brought from differential transport equations for turbulent kinetic energy and turbulent dissipation rate [20] [21] [22] [24]. Shear stress Transport (SST Model) blended the demand of k-w and k-ε models [20, 23, 25].

As the boundary conditions, the characteristics of Jhimruk Hydropower Plant was used, which has a designed output of 12.6 MW with 3 units, each running at 1000 rpm. The net head of the plant is 201.5 m and the total discharge is 7.05 m^3/s. This gives a total flow of 2.35 m^3/s for each unit, according to the design condition. In this study, since a single blade passage was modeled, the mass flow rate for each passage was 97.9 kg/s. However, at off-designed conditions, the guide vanes have to be oriented accordingly to maintain the synchronous speed. The blade, hub and shroud are modeled as no-slip walls such that the fluid velocity near the wall decreases due to wall friction.

The sediment particle was defined as a Quartz material with a density of 2.65 gm/cm^3 and a diameter of 0.1 mm and spherical shape [3] [20]. The particle mass flow rate was based on the amount of ppm of sediment recorded in the site, [4] [16] for the favorable conditions for operation of the turbine, which corresponds to 0.07 kg/s at a concentration of 1000 particles was consider as reference for the simulation [4] [16]. In this study, the model of Tabakoff and Grant [20] was used for modeling the erosion on the blade with Quartz-Aluminum erosion parameters. The erosion rate E from this model is determined from the equation incorporated in ANSYS CFX [4] [20].

This study contains the CFD study in 10 different guide vane opening angles corresponding to different flow conditions. The difference between maximum GV opening and maximum GV closing in rotational degree was found to be 22.8°. This degree was calculated over the axis of the GV rotation. The maximum closing that was used in this study was at 18% opening whereas the maximum opening of 100% was used. The inlet flow used for each of the cases varies according to the GV opening angle. Below 18% opening, it was not possible to use the same mesh quality as the guide vanes remained very close to each other. Hence, 10 guide vane opening angles were used in between 18% and 100%.

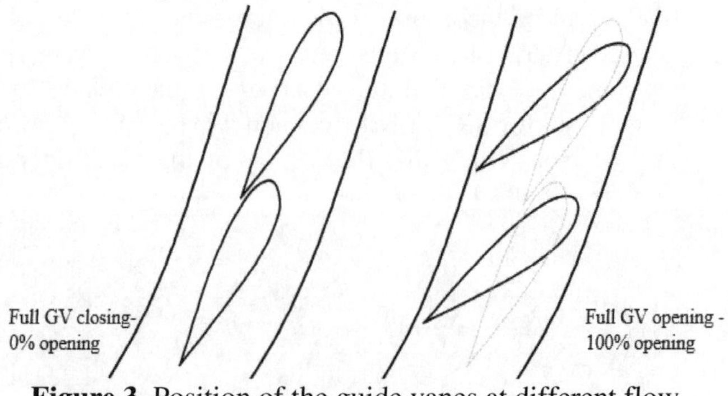

Figure 3. Position of the guide vanes at different flow conditions

Table 3. Guide vane opening in % corresponding to flow conditions

Flow kg/s	132.8	121.2	109.5	97.9	86.2	74.6	62.9	51.2	39.6	27.9
GV open %	100	91	81	72	63	54	45	36	27	18

Apart from the 10 different operating conditions, this study also investigates 5 shapes of the runner blades, which were modified based on the blade angle distribution from inlet to outlet. These shapes were proposed in some of the previous literatures [4] [16] [26], which showed a significant improvement on the erosion for some of the designs. Figure 4 shows the shapes of the blades compared. The blade angle distribution affects the position where the conversion of the hydraulic energy to mechanical energy takes place. Out of these shapes, Shape 3 represents the conventional design. In this case, the blade angle increases linearly from inlet to outlet, i.e. hydraulic energy is converted to mechanical energy linearly. In contrary, Shape 1 converts most of the energy towards inlet of the runner whereas Shape 3 converts towards outlet. Shape 4 and Shape 5 are 'S' shaped blade angles, where the conversion takes place in a more sophisticated way.

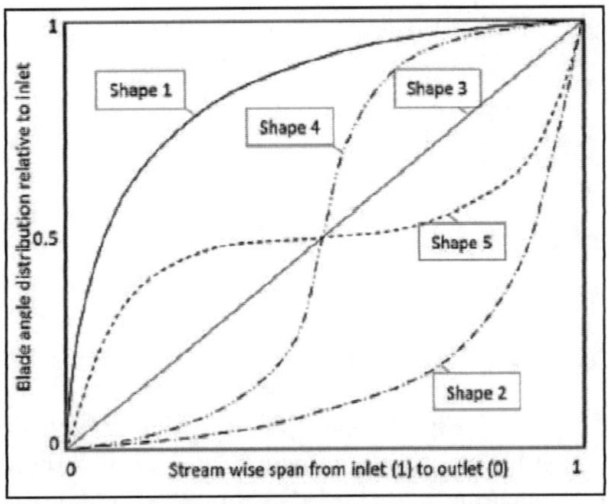

Figure 4. Blade shapes studied [16]

Previous works have compared the erosion potential of these shape of blades at the design condition. However, this study not only focuses on one operating condition, but also develops a trend of how the values of erosion changes in guide vanes and runner blades at 10 operating conditions. As all the simulations were performed for two turbulence models, the study contains results of 100 different combinations.

3. Result and discussion

3.1. Validation of simulated results

The results obtained from both k-ε and SST simulations were first validated with the site measurement values. The absolute velocities obtained at SV Inlet, SV Outlet, GV Inlet, GV Outlet and relative velocities at runner inlet and outlet were compared. The compared results are shown in Figure 6-9.

On studying the absolute velocities versus power, the graphs shown in Figure 6 were obtained, which shows that velocities at corresponding power on k-ε and SST matches, but the experimental values of power output in SV Outlet and GV Inlet are higher than the CFD results of both k-ε and SST. This means that the velocity of water needs to be higher than simulated data for the generation of same power. This is because of position of the sensor placed during the measurement and wear and tear of the turbine runner, guide vane and stay vane. The conditions of the turbine and the sensors locations are presented in Figure 5 [11]. However, the experimental value of power output with respect to the absolute velocity at SV outlet is higher than simulated value of k-ε and SST.

Figure 5. a. Locations of sensors b. Runner condition during experiment c. Guide vane eroded condition

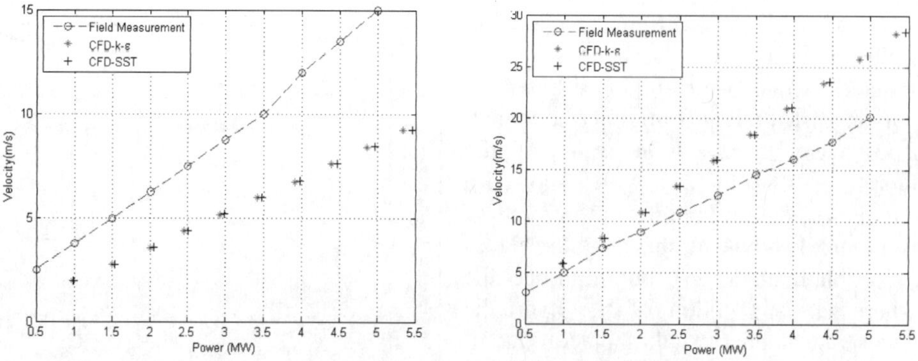

Figure 6. a. Absolute Velocity at SV Inlet b) Absolute Velocity at SV Outlet

Ideally, increasing velocity of water also increases the output power. However, the results of CFD and measurement shows that the trend is deviated from one another while increasing the velocity. In the absolute velocity at GV outlet, power output at 2.5 to 3.5MW are found to be similar, i.e. decreasing trend, whereas for other cases, experimental and simulated values trends are substantially different from one another. This indicates that GV the runner were worn out during the measurement.

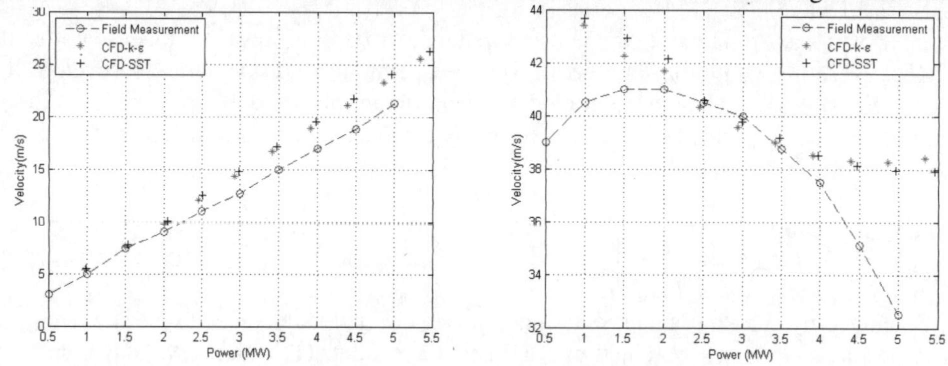

Figure 7. a. Absolute Velocity at GV Inlet b) Absolute Velocity at GV Outlet

When comparing the relative velocities at inlet and outlet, both experimental and simulated values were found to be in a similar trend, i.e. increasing relative velocity gives the increase power output, as shown in Figure 8. However, this trend is slightly deviated from the trend line of relative velocity at runner outlet trend. The relative velocity of outlet shows the similar trend from power output 2MW to 5MW. Normally, the turbines at Jhimruk do not operate at power output less than 2 MW.

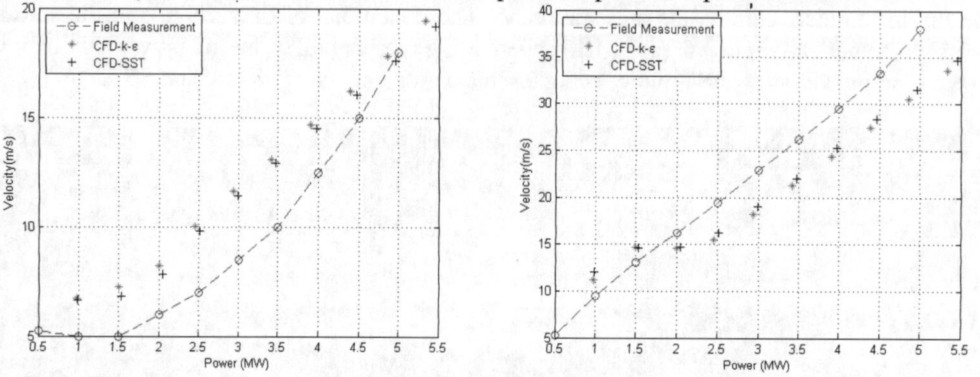

Figure 8. a. Relative Velocity at Runner Inlet b) Relative Velocity at Runner Outlet

Simulation of erosion on the turbine was performed at a specific condition, i.e. spherical shape of the sand with same diameter, concentration and flow rate. In the real turbines, particles have varying shape, size and concentration. To compare erosion between simulated and actual condition, the eroded turbine in the plant was observed and compared with simulated results for extreme cases. Figure 9 a), b) and c) shows the erosion on the runner at BEP, full flow and part flow conditions respectively. It shows that erosion is mostly concentrated at the outlet of the blades for full flow and designed conditions. Turbines in run-off-river type projects such as Jhimruk are mostly vulnerable to full flow conditions. This is because i) the flow has high velocity, which directly influences the erosion and ii) the flow carries plenty of sand particles during monsoon season, when excess water is available. In dry season, the water is cleaner, and is less susceptible to erosion. Figure 9 d) shows erosion on the actual runner, combining all the operating conditions. These observations also show that the runner is heavily eroded on the outlet of the blades. Figure 9 c) also shows that during part flow conditions, erosion is mostly concentration at mid-stream position of blades. Figure 11 shows that the erosion rate would be high in part flow conditions, but in a real scenario, the effect is reduced due to the clean water.

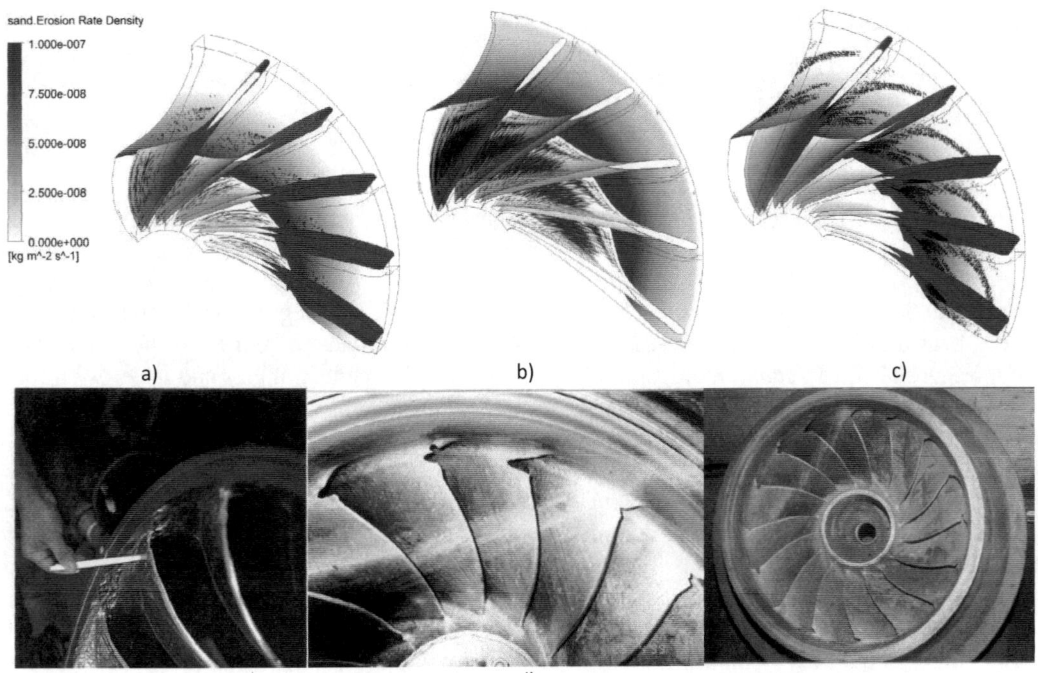

d)

Figure 9. Erosion on the runner at a) BEP b) Full flow condition c) Part flow condition d) Accumulated effect on the actual runner

3.2. Comparison of Efficiency

Figure 10 shows the graph of efficiency vs Guide Vane (GV) opening for five different shape of runner. In this case, the flow conditions are represented by GV opening angles with reference to BEP. Hence, according to Table 4, 0° GV opening angle represents 72% opening. Similarly, 6° opening angle represents 100% opening and -12° represents 18% opening. The best efficiency for all the shapes is obtained between 45-55% of guide vane opening. In this condition, the runner is getting optimum flow and this condition is the design condition for all shapes of runner. When the turbine is operated in part load condition, i.e when the GVs are closing positions, minimum efficiency is obtained due to turbine is getting less flow. Towards the designed condition, the deviation of efficiency for the blades is not very prominent. The best efficiency of 97.71% was obtained in Shape-1 and Shape-3 for 55% GV opening. The deviation is more noticeable at lower and higher GV openings. Towards full opening, Shape-1 and Shape-3 showed best results, whereas towards full closing, these blades had lowest efficiencies. The deviation is more noticeable at lower and higher GV openings. Shape-2 and Shape-4, on contrary, showed best results towards full closing, whereas the efficiencies decreased significantly at larger GV openings. On comparing Shape-5 with other shapes in part and full load conditions, it was seen that this shape has the flattest efficiency curve demonstrating acceptable performances for both part load and full load conditions. The difference is not very

IOP Conf. Series: Journal of Physics: Conf. Series **1042** (2018) 012001 doi:10.1088/1742-6596/1042/1/012001

prominent between the two turbulence models. However, in overall, the efficiency predicted by SST model is higher than k-ε model.

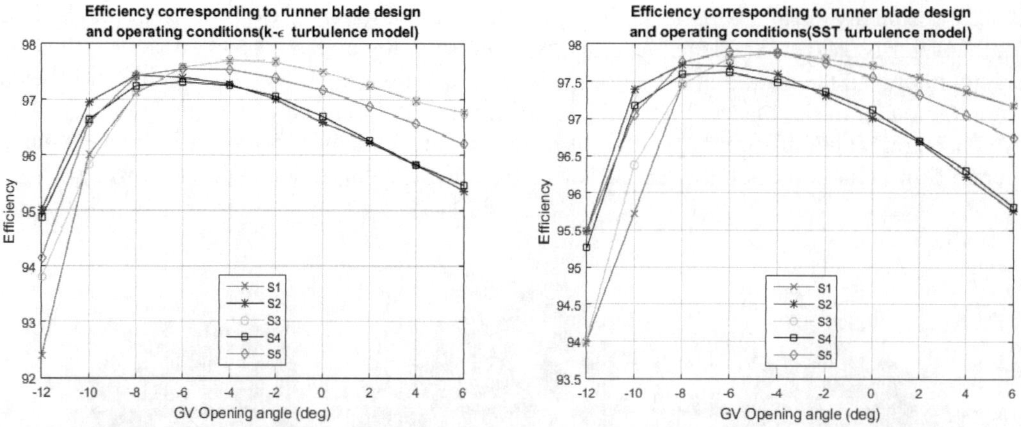

Figure 10. Efficiency corresponding to runner blade design and operating conditions

3.3. Comparison of Erosion

The effect of the change of the runner blade shapes on the erosion of guide vanes and runner was observed. Erosion observed in runner blade and GV for all the designs and turbulence models are shown in Figure 11-12. In general, the trend of the curves of GV opening angles and the erosion for all the shapes is irregular. However, the graph in Figure 11 shows that towards best efficiency points, the erosion in runner blades is minimum. In this condition, particles carried by flow regime is less turbulent and glided with water rather striking towards wall of runner blades. The erosion in the runner blade was maximum towards full GV closing positions. When the guide vane is towards the closing positions, flow is more turbulent and the glided particles with flow experience high centrifugal force causing the particles to strike towards the wall of the runner blade. Out of all the five different blade shapes, Shape-1 and Shape-3 had the maximum erosion at all the GV openings for both the turbulence models. Rest of the blades showed a similar trend towards best efficiencies and full load conditions. Towards full closing, the erosion was found to be maximum for Shape-2 and Shape-4 whereas minimum for Shape-5, as depicted in the figure.

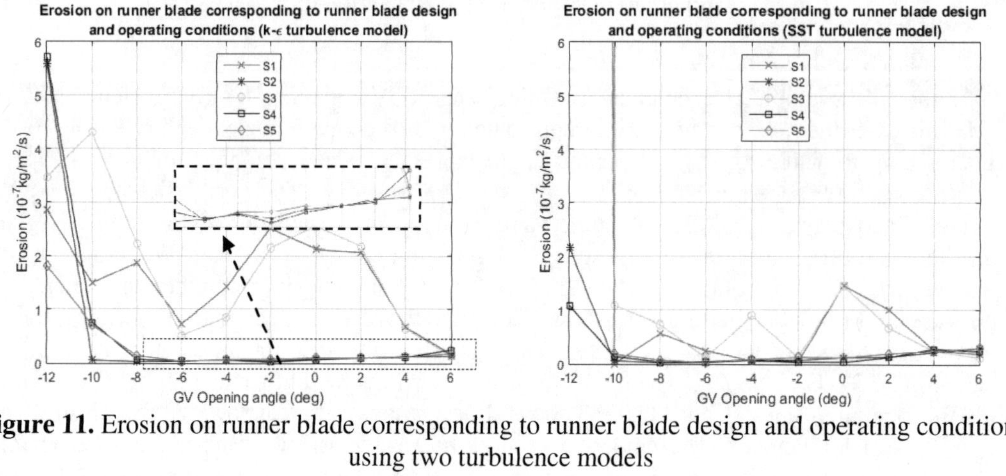

Figure 11. Erosion on runner blade corresponding to runner blade design and operating conditions using two turbulence models

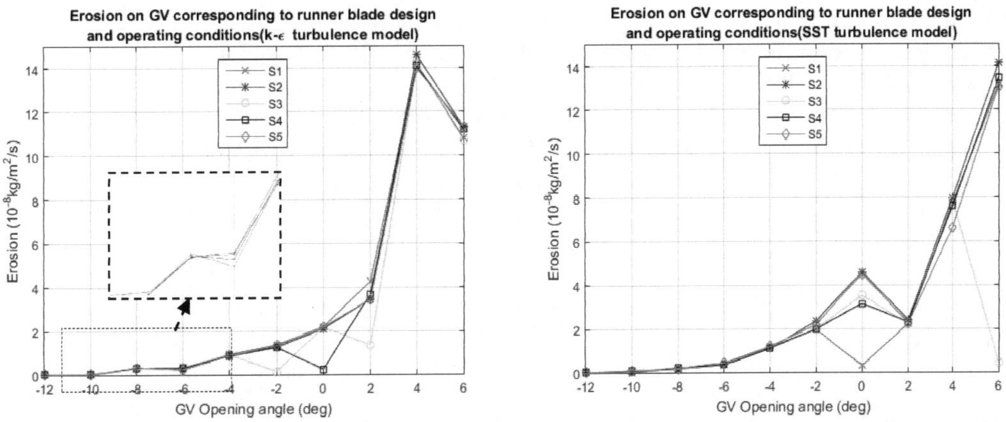

Figure 12. Erosion on guide vane corresponding to runner blade design and operating conditions

4. Conclusion

This study presented the performances of 5 designs of Francis turbine runner blades for 10 operating conditions. The comparison was made based on their potential effects on erosion of turbine components. The results showed that the performance of the turbine changes significantly, depending upon the available flow and eventual guide vane opening angles. The results were also compared with the measurement and observation, done on the same power plant. Following results were obtained from these analyses:

i. Absolute velocities at SV inlet, SV outlet and GV inlet and relative velocities at runner inlet and runner outlet increased in a linear trend with respect to power for both CFD and measurement data. However, absolute velocity at GV outlet follows a nonlinear decreasing trend with respect to the power.
ii. Erosion was mostly concentrated at the outlet of the runner blades for full flow and BEP conditions. At part flow conditions, the erosion from simulation was seen to be concentrated at mid-stream positions. However, comparing that to the real turbine conditions, the effect in part flows are not seen to be vulnerable, because concentration of particles in flow is less in dry season compared to monsoon season.
iii. Taking into account the erosion rate at different operating conditions, distribution of efficiencies and relative velocities for power generated, shape 5 showed the optimum result.

References

[1] Thapa B, 2004, *Sand Erosion of Hydraulic Machinery*, Doctorial Thesis at NTNU.
[2] Rund J J, 2004, *Sediment handling problems Jhimruk Hydroelectric Center Nepal*, Norway: NTNU.
[3] Neopane H P, 2010, *Sediment Erosion in Hydro Turbines*, Doctorial Thesis at NTNU.
[4] Eltvik M, 2013, *Sediment Erosion in Francis Turbines*, Doctoral Thesis at NTNU.
[5] Brekke H, 1984, *The Influence this has on Design Manufacturing Testing and Operation*, General Doctoral Lecture NTH, NTNU, Trondheim.
[6] Dahlhaug O G, Thapa B, 2004, *Sand Erosion in Francis Turbine: A case study from Jhimruk Power Plant*, in IAHR symposium on Hydraulic Machinery and Systems, Stockholm.
[7] Karelin V Y, Denisov A, Wu Y, *Fundamental of Hydroerosive Erosion Theory*, in Hydraulic erosion and corrosion, London, Imperial Collage Press, 20025, pp. 1-52.
[8] Brekke H, Wu Y, Cai B Y, 2002, *Design of Hydraulic Machinery working in sand Laden water, in Series on Hydraulic Machinery- vol 2, Abrasive Erosion & Corrosion of Hydraulic Machinery*, Imperial collage Press, 57 shelton street, pp. 155-186.
[9] Padhy M K, Saini R P, 2007, *A review on silt erosion in hydro turbines*, Elsivier, Renewable and sustainable energy reviews, vol. 12(7), pp. 1974-1987.

[10] Shrestha K P, Thapa B, Dahlhaug O G, Neopane H P, Gurung N, Kayastha A, 2013, *Optimised Desing of Francis Turbine Runner for sand Laden Water*, Hydro Nepal, no. 13, pp. 36-43.
[11] Dahlhaug O G, 2012, *Lecture notes on Design of High Head Francis Turbine*, NTNU, Norway.
[12] Thapa B, Upadhyay P, Dahlhaug O G, Madan T, Ranjil B, 2008, *HVOF coating for erosion resistance of hydraulic turbines: Experience of Kaligandaki-A Hydropower Plant*, in Water Resources and renewable energy Development in Asia, Danang, Central Viatanam.
[13] Brekke H, 2001, *Hydraulic Turbines Design, Erection and Operation,* Norway: NTNU.
[14] Rogner M, 2008, *Design of a high head Francis Turbine that are Exposed for Sand Erosion*, NTNU, Norway.
[15] Meland H, 2010, *A new Design of Francis Turbine in order to reduce sediment Erosion*, Norway: NTNU.
[16] Gjoater K, 2011, *Hydraulic Design of Franics Turbine Exposed to Sediment Erosion*, Master's Thesis, NTNU.
[17] Chitrakar S, 2013, *FSI analysis of Francis Turbine Exposed to Sediment Erosion*, Master Thesis, KTH.
[18] Paulsen J B, 2013, *FSI analysis of a Franics turbine*, Norway: NTNU.
[19] Shrestha K P, Panthe A, Thapa B, Dahlhaug O G, Neopane, H P, Singh T B, 2013, *Design of Model Francis Turbine for Performance test against Sand Erosion*, in 35 IAHR, China, 2013.
[20] A. Inc, 2011 *Ansys CFX solver theory guide*, Release 14, ANSYS.
[21] Nicolisi S, Prisecaru I, Dupleac D, 2013, *Two-phase flow CFD modelling for evaluation of particulate erosion*, U.P.B. Sci. Bull., Series D, vol. 75, no. 1, ISSN 1454-2358.
[22] Gimenez J M, Ramajoa D, Nigroa N M, 2012, *Particle transport in laminar/turbulent flows*, in Mecánica Computacional Vol XXXI,, Salta, Argentina, 13-16.
[23] Mentor F, 1994, *Two-Equation Eddy-Viscosity Turbulence Models for Engineering Applications*, AIAA-Journal 32(8), pp. 269-289, 1994.
[24] Matsubara C, Kuo T, Wu H, 2013, *Comparison of the Effects of k-ϵ, k-ω, and Zero Equation Models on Characterization of Turbulent Permeability of Porous Media*, Journal of Water Resource and Hydraulic Engineering, vol. 2, no. 2, pp. 43-50.
[25] ANSYS, 2006, *Innovative turbulent Modeling: SST Model in ANSYS CFX*, December.
[26] Wei Z, Finstad P H, Olimstad G, Walseth E and Eltvik M, 2009, *High Pressure Hydraulic Machinery, Water Power Laboratory*,, NTNU.

IOP Conf. Series: Journal of Physics: Conf. Series **1042** (2018) 012002 doi:10.1088/1742-6596/1042/1/012002

Literature review relevant to particle erosion in complex geometries

Eirik Volent and Ole Gunnar Dahlhaug

Department of Energy and Process Engineering, Norwegian University of Science and Technology, Høgskoleringen 1, Norway

Email: eirik.volent@ntnu.no

Abstract. Erosion is a challenge in many industries where fluid is transferred through pipe and valve arrangements. Wear can occur in a variety of systems and is often related to the presents of droplets or solid particles in the fluid stream. Solid particles are in many cases present in hydropower systems, and can cause severe damage to system components. Flow conditions, particle size and concentration vary greatly and can thus cause a vast variety of damage, ranging from manageable wear to component failure. The following paper will present a summary of literature relevant to the prediction of erosion in complex geometries. The intention of the review is to investigate the current state of the art, directly relevant to the prediction of wear due to solid particle erosion in complex geometries.

1. Introduction

An understanding of erosion patterns and rates is crucial for designing equipment that can manage challenging erosive flow conditions. Numerical analysis alone can be an adequate tool for predicting erosion rates, however, the quality and accuracy of such methods can be challenging to predict. Experimental methods are a powerful means of calibrating such models for specific geometries and flow conditions, furthermore give more precise predictions of wear and product lifetime.

By understanding the fundamentals of erosion, approaches to validation and application of erosion models, a method may be developed in order to accurately predict erosion rate in specific complex geometries. The following paper will present central literature relevant to identifying the main developments and focus of the current available research related to the prediction of erosion in complex geometries.

2. Mechanism of erosion

The erosion phenomenon is widely studied and several theories have been produced in order to describe the mechanisms that occur as particles impact and remove material from a surface. Based on the study of free moving particles and particle deformation upon impact, Bitter [1],[2], proposed a two-part representation describing the modes of wear. *Deformation wear*, W_d ,wear on surface due to particles impacting normal to the surface and *Cutting wear,* W_{C1}, W_{C2} , due to particles impacting at an acute angle relative to surface. Deformation wear and cutting wear occurs, in most cases, simultaneously. It was identified that the cutting wear, is the dominating erosion in ductile materials, whereas deformation wear, is the dominating erosion in brittle materials. This is presented graphically in Figure 1 and Figure 2. According to Bitter [2], the plastic deformation of the surface occurs when the elastic limit of the surface material is reached as a result of repeated particle impact. In this process the surface is deformed plastically and work hardened. As the area cannot deform any further the surface cracks and small fragments are detached. As particles impact on a surface at an acute angle, cutting wear can occur due to shear stress of the surface material being exceeded. Particle velocities are divided into two components, vertical and horizontal, relative to the surface. When velocity normal to the surface is large enough to plastically deform the surface layer, cutting and scratching can occur due to the addition of the horizontal particle velocity component. Bitter [1] divides cutting wear further into two equations dependent upon the velocity of the particle after impact. When horizontal velocity of the particle is present post impact, the erosion is represented by W_{C1}. When the particle has no horizontal velocity after impact the erosion is represented by W_{C2}. The total wear, W_t, is defined as the sum of deformation wear and cutting wear [2], [1].

$$W_D=\frac{1}{2}\cdot\frac{M\cdot(V\cdot\sin(\alpha)-K)^2}{\varepsilon} \tag{1}$$

$$W_{C1}=\left(\frac{2\cdot M\cdot C(V\cdot\sin(\alpha)-K)^2}{(V\cdot\sin(\alpha))^{1/2}}\right)\left(V\cdot\cos(\alpha)-\frac{(C\cdot(V\cdot\sin(\alpha)-K)^2}{(V\cdot\sin(\alpha))^{1/2}}\cdot e\right) \tag{2}$$

$$W_{C2}=\frac{\frac{1}{2}\cdot M\cdot\left[V^2\cdot\cos^2(\alpha)-K_1\left(V\cdot\sin(\alpha)-K^{\frac{3}{2}}\right)\right]}{e} \tag{3}$$

$$W_t=W_D+W_{c1}\text{ or }W_D+W_{c2} \tag{4}$$

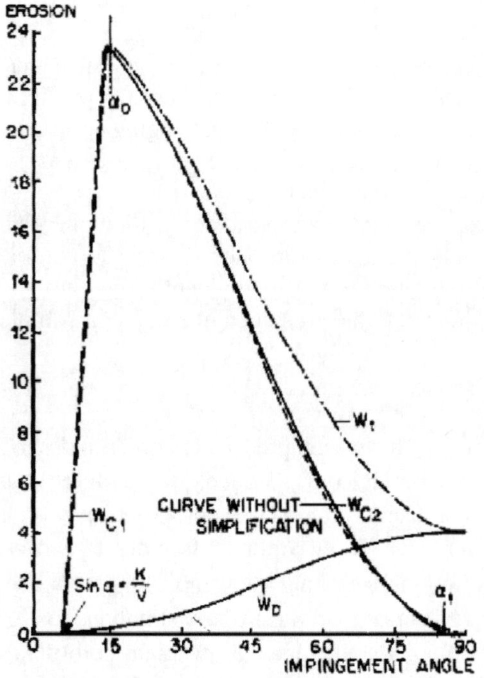

Figure 1. Characteristic erosion of ductile material [1].

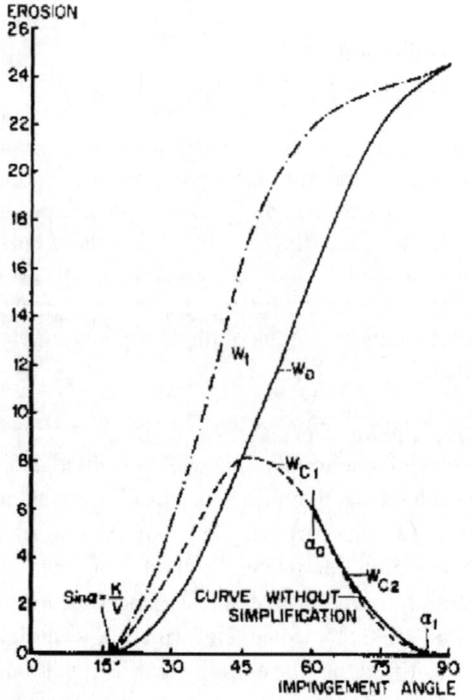

Figure 2. Characteristic erosion of brittle material [1].

IOP Conf. Series: Journal of Physics: Conf. Series **1042** (2018) 012002 doi:10.1088/1742-6596/1042/1/012002

Finnie [3] proposed that erosion of a ductile surface material, due to solid particle impact in a fluid stream, is dependent on the fluid flow and on the mode of material removal. Erosion of ductile and brittle materials were identified as diverse mechanisms and only a model for erosion of ductile materials was considered.

Finnie introduced a model describing the removal of material as a result of micro-cutting of the surface. Based on micro-geometry of an impacting particle on surface, and the impingent angel, Finnie proposed two equations. The two equations, (5) and (6), present a prediction of the volume of removed material (Q) as a function of a single particle impact. The two erosion modes are dependent on the velocity of the particle after cutting.

$$Q=\frac{M\cdot V^2}{\rho\cdot\varphi\cdot K_2}\left(\sin(2\cdot\alpha)-\frac{6}{K_2}\cdot\sin^2(\alpha)\right)\ \text{If, }\tan(\alpha)\ 0\leq\frac{K_2}{6} \tag{5}$$

$$Q=\frac{M\cdot V^2}{\rho\cdot\varphi\cdot K_2}\left(\frac{K_2\cdot\cos^2(\alpha)}{6}\right)\ \text{If, }\tan(\alpha)\geq 0\frac{K_2}{6} \tag{6}$$

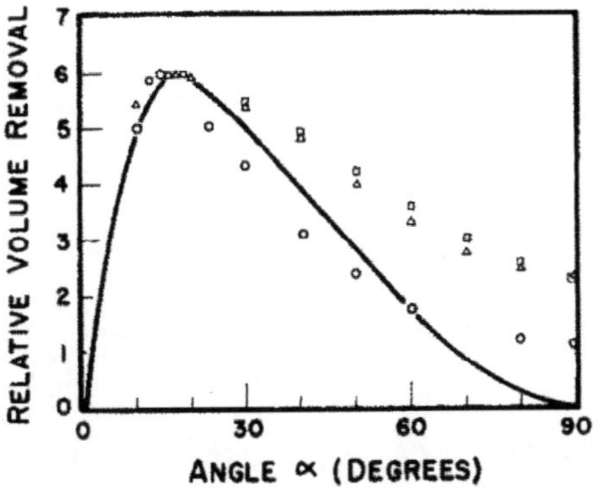

Figure 3. Relative volume removal vs particle impingement angle as presented by Finnie [4]. Predicted wear is represented by continues curve, experimental data for different ductile materials are represented as points

Finnie [4] based much of his study on realistic fluid flow erosion issues. Thus, considering the fluid flow aspect and particle shape in depth. Finnie's early model underpredicts the erosion at high impingement angles as can be seen in Figure 3. Later Finnie [3] introduced a modified model to address the issue of wear at high impingement angles. Additionally, identifying several parameters that may affect the erosion, and concluding that not all parameters can be measured or controlled, furthermore not being implemented in a model.

Tilly [5] introduced the terms primary and secondary erosion, which is caused by the first impact and fragmentation of the particle respectively. Particle impact angle and erosion rates for primary and secondary erosion is presented in Figure 5. Addressing the mechanism occurring at high impingement angles has shown to be challenging to predict analytically. Tilly [5] presents a theory that introduces an erosion mechanism of ductile materials at large particle impingement angle. In circumstances where particle impact angle is high, the particle can fracture upon impact and the fragments can wear the surface surrounding the larger impact area. Figure 4 show a microscope image where the primary and secondary damage can be observed.

IOP Conf. Series: Journal of Physics: Conf. Series **1042** (2018) 012002 doi:10.1088/1742-6596/1042/1/012002

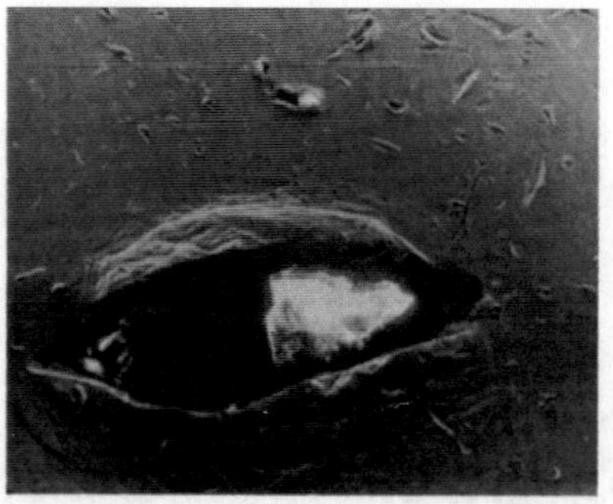

Figure 4. Primary indentation surrounded by secondary damage produces by 700 μm quartz particle against H 46 steel at 800 ft.sec^{-1} (x200). [5].

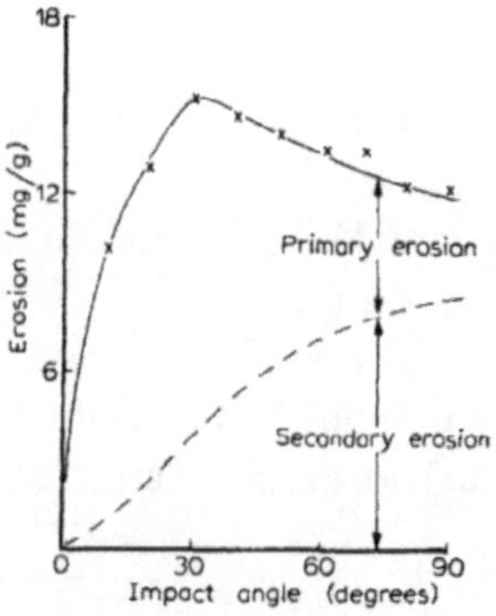

Figure 5. Ductile erosion curve. The influence of impact angle for 135 μm quartz against H 46 steel at 1200 ft.sec^{-1}. [5].

Tilly [5] derived his erosion model from the equation of motion and kinetic energy. The two stage mechanism is represented by the following equations.

$$\varepsilon_1 = \hat{\varepsilon}_1 \cdot \left(\frac{V}{V_{ref}}\right)^2 \cdot \left[1 - \left(\frac{d_0}{D}\right)^{\frac{3}{2}} \cdot \left(\frac{V_0}{V}\right)\right]^2 \tag{7}$$

$$\varepsilon_2 = \hat{\varepsilon}_2 \cdot \left(\frac{V}{V_{ref}}\right)^2 \cdot F_{d,v} \tag{8}$$

$$\varepsilon_{total} = \varepsilon_1 + \varepsilon_2 \tag{9}$$

Further, Tilly[5] identified that the wear was dependent on the particle size. For particle size larger than 100μm the relationship become linear, hence the ratio of mass erode material / mass erosive material becomes constant. This relationship is shown in Figure 6.

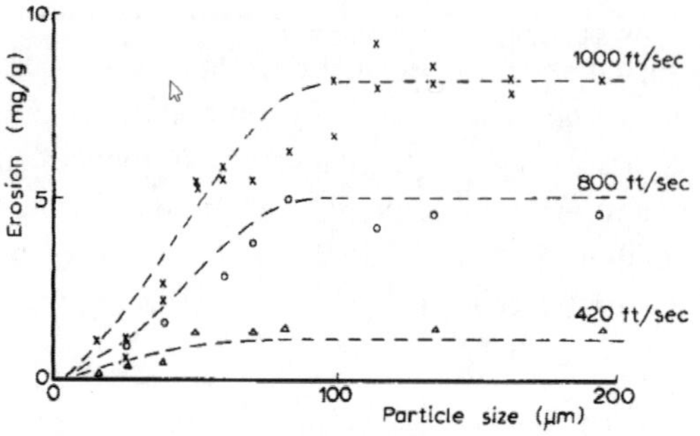

Figure 6. Erosion ratio versus particle size and particle impact velocity [5]. Curves derived from equation (9), points from experiment data.

Tabakoff and Grant [6] proposed a semi empirical model for predicting erosion. In the presented model the erosion rate is dependent on the particle impingement angle and velocity. In addition, several constants related to the particle-surface interaction was incorporated. Tangential particle velocity post impact is taken into account as a restitution factor. It was assumed that the erosion occurs in two stages dependent on the impingement angle and that the models is representative for both small and large particle impingement angle as well as a combination. The Tabakoff and Grant model allows for great adjustment to specific erosion challenges. However, the accuracy of the model strongly depends on the determination of a variety of constant that is required.

$$E=k_7 \cdot f(\alpha) \cdot V^2 \cdot Cos^2(\alpha) \cdot \left[1-R_T^2\right]+f(V_{PN}) \tag{10}$$

$$f(\alpha)=\left[1+k_5 \cdot k_6 \cdot \sin\left(\alpha \cdot \frac{\frac{\pi}{2}}{\alpha_0}\right)\right]^2 \tag{11}$$

$$R_t=1-k_4 \cdot V \cdot \sin(\alpha) \tag{12}$$

$$f(V_{PN})=k_3 \cdot (V \cdot \sin(\alpha))^4 \tag{13}$$

$$k_5 \begin{Bmatrix} \begin{cases} 1.0, & \text{if} \quad \alpha \leq 2\alpha_0 \\ 0.0, & \text{if} \quad \alpha > 2\alpha_0 \end{cases} \end{Bmatrix} \tag{14}$$

3. Erosion in valves and complex geometries

Later literature introduces models that are developed in correlation with modern techniques such as computational fluid dynamics simulations (CFD).

DNV recommended practice 0501 [7], present a guideline for material selection and erosion evaluation for a selection of pipe geometries. The general equation allows for great adjustability to different factors related to materials and particles. The correlation between erosion rate E, particle mass m, velocity U and surface material function / angle is given by:

$$\dot{E} \sim \dot{m}_p \cdot K_8 \cdot V^n \cdot F(\alpha) \tag{15}$$

$$\dot{E}_l=\frac{(\dot{m}_p \cdot K_9 \cdot V^n \cdot F(\alpha))}{\rho_t \cdot A_t} \cdot C_{Unit} \tag{16}$$

$$F(\alpha)=\sum (-1)^{(i+1)} \cdot A_i \cdot \left(\frac{\alpha \cdot \pi}{180}\right)^i \tag{17}$$

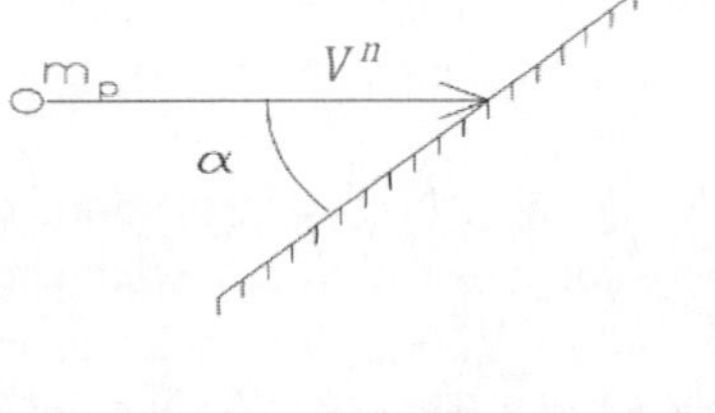

Figure 7. Parameters characterising erosion impact on a surface, ref. equation (15), [7].

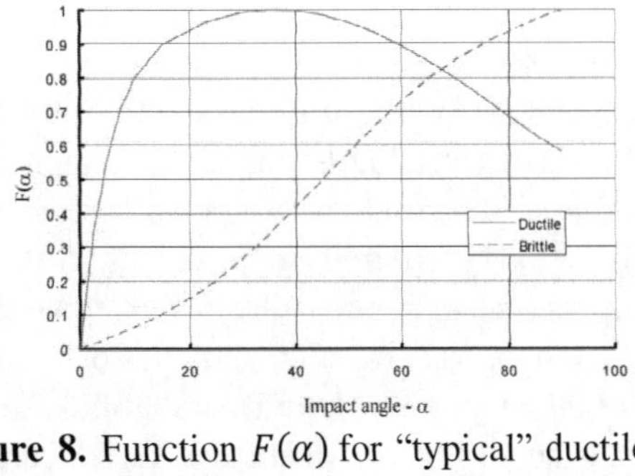

Figure 8. Function $F(\alpha)$ for "typical" ductile and brittle materials [7].

The DNV model correlates with Haugen's work [8] and was developed by the use of CFD analysis in parallel with experimental tests. Haugen introduced experimental data for a wide range of materials as presented in Figure 9. The barchart show a comparison of erosion rate of different materials such as ceramics and hard metals, extensively used in industry, relative to steel.

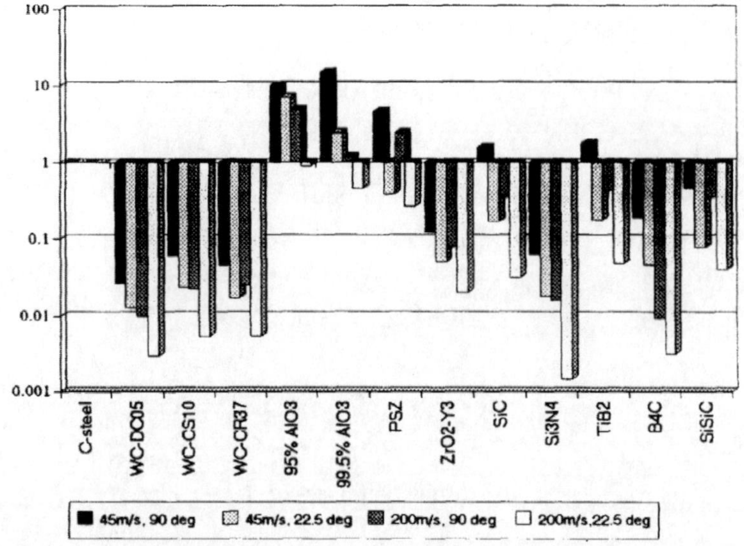

Figure 9. Relative erosion resistance of solid WC and ceramics relative to C-steel. [8]

One of the more recognized CFD software used for erosion prediction in fluid systems is Ansys and Fluent. These software packages enables many models to be implemented, Finnie [4] and Tillly [5] to mention some. However, the default model used in the Fluent code as shown in equation (18), is similar to equation (15), [9]. The model display similarities with the DNV Haugen model [8]. However, with the distinction of a particle size function, d_f. This allows for a more realistic representation of particle distribution, as the size and number of impinging particles can vary vastly in real life events.

$$\dot{E}= \sum_{P=1}^{N_{particles}} \frac{\dot{m}_p \cdot C(d_f) \cdot f(\alpha) \cdot V^{b(v)}}{A_{face}} \tag{18}$$

McLaury[10] presents a three-part investigation by the means of CFD in the attempt to create a comprehensive method for erosion rate estimation for valve designers in industry. The CFD method is described in short as follows: 1. Solving the Navier-Stokes equation, and turbulence model, 2. Applying the Euler-Lagrangian method for particle trajectory tracing, and 3. Applying an erosion model to estimate the erosion rate.

Erosion model and constants used for the test was based on Finnie [4] and Bitters [1] work. By comparing numeric predictions and experimental data, a relatively large discrepancy was found. A proposed reason for the difference in results was related to the turbulence as calculated in the computational dataset. Furthermore, the change in geometry as a function of erosion was not considered in the CFD analysis resulting in a potential inaccuracy of the prediction.

Zhang [11] used similar approach as McLaury [10] by implementing a three stage method. Results from the experimental setup with erosion probes and laser Doppler velocimetry was compared with CFD predictions. A nozzle projecting particles on to an erosion probe (and LDV field) in a tank was used as experimental setup. It was found that the velocities and particle trajectories had good correlation. Zhang claim that the CFD based method as described is a sufficient way to predict erosion, especially in complex geometries.

Mansouri [12] developed a methodology for predicting erosion and implement relevant variables into an equation set that was applied in a CFD code. The aim was to predict wear and erosion material wastage with a high accuracy by combining CFD analysis and empirical data. The three stage technique as presented by Mansouri is in short described as following: 1. Measure erosion depth on a specimen via 3D profilometry, 2. Calculate the particle speed, angle and frequency of impact as a function of radial position via CFD simulation and 3. Fit an equation which correlates the particle impact information to thickness loss (erosion depth). The particle flow was investigated by the means of particle image velocimetry (PIV) measurement technique to quantitatively determine particle distribution and velocity. PIV, CFD and erosion test-parameters was combined with an erosion prediction equation which produced a profile that was compared with experimental data. A difference of less than 14% between prediction model and test data was found.

Fordere [13] presents a numerical study of a complex needle – seat control valve based on parameters representative for typical oil and gas fields. The CFD based approach is similarly defined as in Zhang [11] and McLaury[10]. It is argued that the standard erosion equation is not satisfactory for complex geometries such as choke valves. A design philosophy at which impingement angles and velocities in the diverging section of the choke are reduced, is applied and tested with success [13]. Comprehensive testing was completed in order to validate results, however, a limited amount of results was presented.

Wallace [14] presents an in-depth study on prediction of wear in complex geometries such as control valves. Experimental erosion parameters were obtained by jet type erosion tests. Three stages of modelling were discussed: flow field, particle dynamics and wear models. The models used in this study show that CFD could predict the flow characteristics of complex geometries. It can also be concluded that the location of erosion is well predicted by CFD, however, this study demonstrates that the modelling underpredicts (by 10-15 times) erosion compared with experimental values. For complex geometries such as choke valves used in the study, a significant difference between prediction and measured values was found.

The study presented by Nøkleberg and Søntvedt [15] was focused towards lifetime prediction of valves to reduce cost related to maintenance. Nøkleberg and Søntvedt present a comparison of a numerical study and experimental tests of a choke valve for oil and gas applications, Figure 10. The oil and gas sector is an industry with challenging conditions such as high flow velocities, erosive – corrosive fluids, extreme pressures and temperature. Erosion by droplets was considered to be a major contributor to the erosion of choke valves. Droplets can often occur in two or multiphase flow, and it can be difficult to establish whether observed erosion is due to solid particles or by droplets.

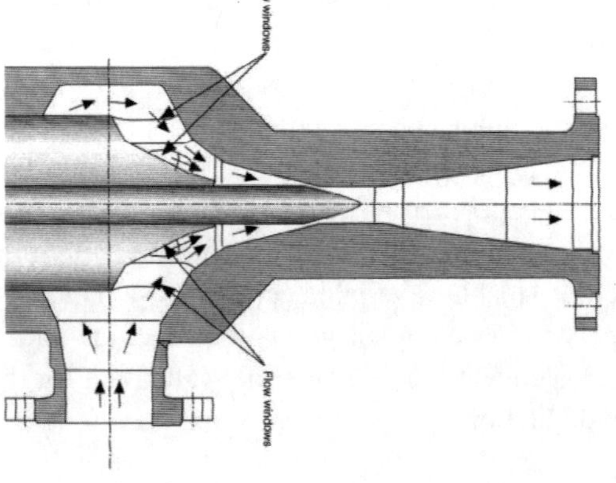

Figure 10. Modified needle-seat choke valve [16]

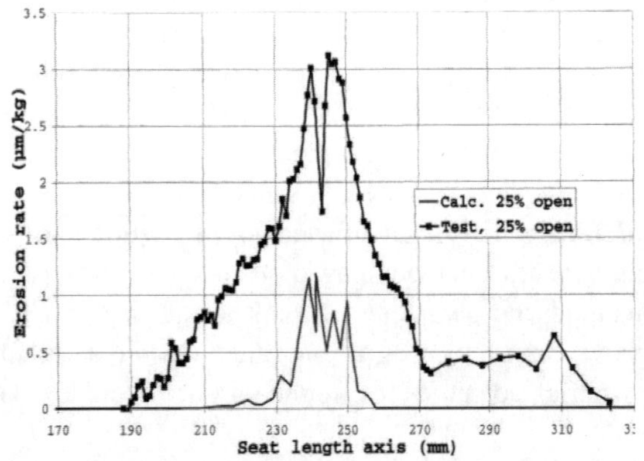

Figure 11. Maximum erosion on steel seat and of modified needle and seat Choke, 25% open [16].

The erosion model and CFD correlated well with field studies and the results was used to develop a modified choke design. The modified design, Figure 10, introduces a reduction in particle impingement angles, to optimize the lifetime of the equipment. Nøkleberg and Søntvedt [16] made a quantitative

IOP Conf. Series: Journal of Physics: Conf. Series **1042** (2018) 012002 doi:10.1088/1742-6596/1042/1/012002

comparison in their modelling of a modified needle and seat choke, and showed that the technique (using the structured Fluent code was able to adequately predict wear location.

In later work Nøkleberg and Søntvedt [16] compared numerical models and experimental data, considering erosion by solid particles only. The tests were performed on the modified choke valve at DNV flow loop which was purpose built for the application. The experiment was conducted with supersonic velocities that reflect real life operating conditions for many choke valves. It was found that the erosion model and parameters used for the application underestimated the real erosion depth magnitude (by 2-3 times) as shown in Figure 11. By comparing wear in an optimized needle-seat valve with an external sleeve valve it was found that the erosion rate reduction of the needle-seat valve was as much as 50-100 times.

Wang [17] presents a method of predicting wear in a throttle valve, which is prone to erosion due to sand slurry flow. A CFD analysis was made and compared with empirical results. The study revolves around a specific complex valve geometry. The aim of the study was to validate a prediction method, parameters, and to optimized design based on modelling. A good correlation between test and simulation was found, however, it was identified that the erosion prediction and experimental results had a diverging correlation with time. Change in surface geometry as a function of erosion, which is not considered by the CFD analysis, was considered to be the main contributing factor of the discrepancy.

4. Discussion

Predicting erosion accurately is a challenge, and many models have been made to account for variables that effect the particle – surface interaction in the erosion process. Meng and Ludeman [18] introduced an overview of 28 different erosion models accounting for up to 34 variables. The strive to derive models that represent erosion accurately have led to a wide range of models that allow for vast range of parameters. [18]. However, the most relevant parameters contributing to erosion is identified as the particle velocity, mass, impingement angle and surface material properties.

Bitter, Finnie and Tilly among others, produces some of the most recognized models for predicting erosion by solid particles. Later literature introduces models and methods that are developed in correlation with modern techniques such as CFD.

The particle trajectories and impingement statistics is in present methods commonly done by CFD simulation, and the erosion models are generally applied to calculate the erosion based on the particle – surface interaction only. This method has been presented in available literature and has good predictive capabilities with respect to erosion location. However, it is the belief that with the growing availability of CPU power, the accuracy of both erosion location and magnitude can be studied in detail.

The importance of precise test results in addition to accurate CFD models has shown to be critical in the prediction of wear. When evaluating erosion in complex geometries and determining component lifetime a high level of accuracy is required. Based on a "comprehensive literature review of solid particle erosion modelling in oil and gas", Paris [6] concluded that the available generic erosion models, method and parameters are not adequate for predicting wear in complex geometries such as choke valves. In such cases, it is required to do experimental erosion tests and CFD model validation for the specific geometries in order to produce accurate erosion predictions.

5. Conclusion, discussion and further work

The current methods for erosion modelling and wear prediction is found to be sufficient for establishing the location of erosion. However, the magnitude and rate of erosion is often underpredicted. This is challenging with respect to product lifetime prediction for erosion prone components. This applies particularly to complex geometries such as control valves.

Based on the available information, it is concluded that in order to investigate the discrepancy between numerical model prediction and experimental data for erosion rates, the fluid flow has to be analysed for specific geometries. When detailed numerical analysis can be compared with experimental data, it is believed that factors adding to discrepancy between results may be identified. Furthermore, a

IOP Conf. Series: Journal of Physics: Conf. Series **1042** (2018) 012002 doi:10.1088/1742-6596/1042/1/012002

justified correction can be made to the models or parameters in order to achieve a more accurate erosion prediction for the specific geometry.

In order to better understand the erosion rates for a specific complex geometry, the primary step is to investigate the flow characteristics, and how these characteristics are represented by different numerical parameters and simplifications. When a numerical dataset is obtained, laboratory studies is required in order to validate numerical parameters. Upon the comparison and potential correction of numerical models and the specific flow structure, the particle trajectories can be calculated, measured and further used for accurate prediction of erosion rate in the specific geometry.

Table 1. Nomenclature.

Parameter	Description	Unit
A_t	Area exposed to erosion	[m^2]
A_i	Dimensionless parameter group	
A_{face}	Cell face area	[m^2]
$b(v)$	Function of relative particle velocity	
$C(d_f)$	Particle distribution function	
C_{Unit}	Unit conversion factor	
d_0	Threshold particle diameter below which no erosion occurs	[μm]
D	Particle diameter	[μm]
e	Cutting wear factor	[g f cm / sec^3]
E_p	Erosion parameter, ratio of metal penetration and mass of sand hitting the material	[mm / kg]
$\dot{E}$	Erosion rate	[mm / year]
E	Erosion rate, mass loss / impact mass	
$F_{d,v}$	Fragmentation	
$F(\alpha)$	Material function	
K	Velocity at with plastic deformation occurs	
K_{1-7}	Constant	[cm / sec]
K_9	Material constant	[(m/s) ^-n]
$\dot{m}_p$	Mass flow rate of sand	[kg / s]
M	Total mass of impinging particle	[g f sec^2 /cm]
ρ_t	Target material density	[kg / m^3]
V	Particle velocity prior to impact	[m/s]
V_0	Threshold particle velocity below which no erosion occurs	[m / s]
V_{ref}	Reference velocity	[m / s]
W_D	Wear due to repeated deformation impact	[cm^3]
W_{C1}	Deformation wear	[cm^3]
W_{C2}	Deformation wear	[cm^3]
α	Particle impact angle	
ε	Deformation wear factor, energy required to remove a unit volume of surface material by deformation of scratching	[g f cm / sec^3]
ε_1	Primary erosion rate	
ε_2	Primary erosion rate	
φ	Ratio of depth of contact to depth of cut	

References

[1] Bitter J G A 1963 A study of erosion phenomena II *Wear* **6** 169–90
[2] Bitter J G A 1963 A study of erosion phenomena part I *Wear* **6** 5–21
[3] Finnie I 1972 Some observations on the erosion of ductile metals *Wear* **19** 81–90
[4] Finnie I 1960 Erosion of surfaces by solid particles *Wear* **3** 87–103
[5] Tilly G P 1973 A two stage mechanism of ductile erosion *Wear* **23** 87–96
[6] Parsi M, Najmi K, Najafifard F, Hassani S, McLaury B S and Shirazi S A 2014 A comprehensive review of solid particle erosion modeling for oil and gas wells and pipelines applications *J. Nat. Gas Sci. Eng.* **21** 850–73
[7] Veritas D N 2007 Recommended Practice RP O501 Erosive Wear in Piping Systems *DNV Recomm. Pract.* **4**
[8] Haugen K, Kvernvold O, Ronold A and Sandberg R 1995 8th International Conference on Erosion by Liquid and Solid ImpactSand erosion of wear-resistant materials: Erosion in choke valves *Wear* **186** 179–88
[9] Anon FLUENT 6.3 User's Guide
[10] McLaury B S, Shirazi S A and Burden T L 2000 Effect of Entrance Shape on Erosion in the Throat of Chokes *J. Energy Resour. Technol.* **122** 198–204
[11] Zhang Y, Reuterfors E P, McLaury B S, Shirazi S A and Rybicki E F 2007 Comparison of computed and measured particle velocities and erosion in water and air flows *Wear* **263** 330–8
[12] Mansouri A, Arabnejad H, Shirazi S A and McLaury B S 2015 A combined CFD/experimental methodology for erosion prediction *Wear* **332–333** 1090–7
[13] Forder A, Thew M and Harrison D 1998 A numerical investigation of solid particle erosion experienced within oilfield control valves *Wear* **216** 184–93
[14] Wallace M S, Dempster W M, Scanlon T, Peters J and McCulloch S 2004 Prediction of impact erosion in valve geometries *Wear* **256** 927–36
[15] Nøkleberg L and Søntvedt T 1995 8th International Conference on Erosion by Liquid and Solid ImpactErosion in choke valves—oil and gas industry applications *Wear* **186** 401–12
[16] Nøkleberg L and Søntvedt T 1998 Erosion of oil&gas industry choke valves using computational fluid dynamics and experiment *Int. J. Heat Fluid Flow* **19** 636–43
[17] Wang G R, Chu F, Tao S y, Jiang L and Zhu H 2015 Optimization design for throttle valve of managed pressure drilling based on CFD erosion simulation and response surface methodology *Wear* **338–339** 114–21
[18] Meng H C and Ludema K C 1995 Wear models and predictive equations: their form and content *Wear* **181** 443–57

IOP Conf. Series: Journal of Physics: Conf. Series **1042** (2018) 012003 doi:10.1088/1742-6596/1042/1/012003

Simplified hydrodynamic analysis on the general shape of the hill charts of Francis turbines using shroud-streamline modeling

I Iliev[1], C Trivedi[1] and O G Dahlhaug[1]

[1*]Waterpower Laboratory, Department of Energy and Process Engineering, Norwegian University of Science and Technology, NO-7491 Trondheim, Norway

E-mail: igor.iliev@ntnu.no

Abstract. The paper presents a simplified one-dimensional calculation of the efficiency hill-chart for Francis turbines, based on the velocity triangles at the inlet and outlet of the runner's blade. Calculation is done for one streamline, namely the shroud streamline in the meridional section, where an efficiency model is established and iteratively approximated in order to satisfy the Euler equation for turbomachines at a wide operating range around the best efficiency point (BEP). Using the presented method, hill charts are calculated for one splitter-bladed Francis turbine runner and one Reversible Pump-Turbine (RPT) runner operated in the turbine mode. Both turbines have similar and relatively low specific speeds of $n_{sQ} = 23.3$ and $n_{sQ} = 27$, equal inlet and outlet diameters and are designed to fit in the same turbine rig for laboratory measurements (i.e. spiral casing and draft tube are the same). Calculated hill charts are compared against performance data obtained experimentally from model tests according to IEC standards for both turbines. Good agreement between theoretical and experimental results is observed when comparing the shapes of the efficiency contours in the hill-charts. The simplified analysis identifies the design parameters that defines the general shape and inclination of the turbine's hill charts and, with some additional improvements in the loss models used, it can be used for quick assessment of the performance at off-design conditions during the design process of hydraulic turbines.

1. Introduction

Efficiency of hydraulic Francis turbines at off-design operating conditions is governed by the shapes of their performance hill-charts [1-3]. All parameters that define the wetted surfaces of a turbine passage, both rotating and stationary, has influence on the total efficiency at each operating point defined by the head, discharge and rotational speed of the runner [4]. Depending on the type and location, turbine losses can be typically classified into

- Spiral casing losses (due to skin friction and secondary flow);
- Stay and guide-vane losses (due to secondary flow, wake mixing, skin friction and incidence);
- Runner losses (due to flow incidence at inlet, skin friction in the blade channels, three-dimensional and curvature effects and residual swirl at the outlet);
- Draft tube losses (due to skin friction, flow in an elbow and channel divergence).

Not listed in the losses classification is the slip effect in the runner, which is mainly governed by the (1) relative whirl in the blade channel generated by the finite number of runner blades, and (2) the energy distribution along the blade length. These effects are usually modelled with a slip factor that has indirect influence on the peak efficiency and the location of the BEP point in the hill-chart.

Experimental and numerical results have showed that runner and draft-tube losses can account for up to 50% of the total losses at best-efficiency point (BEP) [5-7]. Normally, runner and draft-tube losses will further increase to an even higher percentage when the turbine is being operated at off-design operating conditions, i.e. part-load (PL) and full load (FL). Considering the global parameters of the turbine, prior knowledge of the shape of the hill-chart can be obtained by conducting a single-streamline hydrodynamic performance analysis. Mean-streamline models are usually used for performance prediction in the design stage of both compressible and incompressible flow in runners and impellers of mixed-flow turbo machines [8-11]. The accuracy of these models, however, relies heavily on the needed coefficients that are rarely reported in the open literature.

On the other hand, Computational Fluid Dynamics (CFD) is a powerful tool that can predict the hill-charts with a high accuracy and is becoming more and more practical for automatically guided design and optimization of turbomachinery [12-14]. However, when analytical insight is needed in order to provide understanding of what drives the general shape of the turbine hill charts, CFD is usually not the way to go.

The focus in this paper is to examine the ability to predict the general shape of the hill-chart of a two low-specific-speed Francis turbines by only using simplified models for the runner losses based on the velocity triangles at the inlet and outlet of the shroud streamline. Since the only difference between the two examined turbines is in the runner design, it indicates that the differences in the general shapes of their hill charts (experimentally measured) could be linked with the losses in and downstream of their runners.

2. Specifications and description of the analyzed runners

Two low specific speed Francis turbine runners, which were available for model tests at the Waterpower laboratory, were used for the theoretical analysis of the general shape of their hill-charts. Both runners are designed to fit in the same test-rig, i.e. same distributor and draft-tube, but originate from different research activities in the aforementioned laboratory [15, 16]. Marked as "F99" is the splitter-bladed Francis turbine runner, while "RPT" represents the reversible pump-turbine runner (see figure 1 below). Global parameters that were used in the theoretical analysis are specified in table 1.

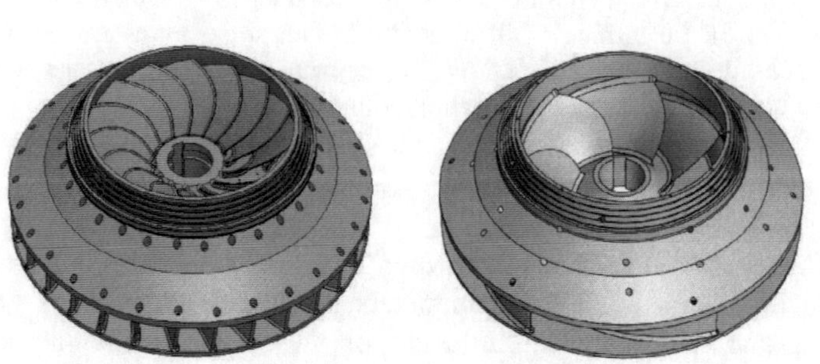

Figure 1. Comparison of the geometries for both runners.
F99 – on the left, RPT – on the right.

The F99 runner has 15 full-length blades and 15 splitter blades with a model peak efficiency of $\eta_h = 92.6\%$ when operated in an open loop under a net head of $H_n = 12\ [m]$. The RPT runner has 6 full-length blades with a model peak efficiency of $\eta_h = 89.6\%$ when operated in the same rig as the F99 model. The common distributor is comprised of 14 stay vanes and 28 guide vanes and the common draft-tube is an elbow type. Hence, the most effective difference between both runners is the reaction ratio ($R^{F99} = 0.533$ and $R^{RPT} = 0.789$) and the prescribed stream wise angle distribution along the blade's mid surface. The specific speeds of both runners are $n_{sQ} = 23.3$ and $n_{sQ} = 27$ in $[rpm\ m^{3/4}\ s^{-1/2}]$, for F99 and RPT respectively.

Table 1. Global parameters used as input for both runners

	F99		RPT	
Parameter	Inlet	Outlet	Inlet	Outlet
Diameter D [mm]	621.6	349	630.5	349
Runner width b [mm]	58.7	/	58.7	/
Blade angle β [deg]	63.38	19.62	12.02	13.45
Number of blades Z [-]	30	15	6	6
Blade length L_b [mm]	250		690	

Experimental data used for comparison is obtained from model tests done at the Waterpower Laboratory from different measuring campaigns. The test rig has the possibility to be operated in open and closed loop and according to the IEC 60193 standard for model tests [17].

3. Description of the solution procedure and modelling of the runner losses

The calculation of efficiency for the operating points used to construct the hill charts is done by modelling the incidence and residual swirl losses in the runner, as well as the friction losses in the runner and draft tube. These losses are modelled as a function of vector components from the inlet and outlet velocity triangles on the shroud streamline of the runner. Iterative procedure is than constructed using MATLAB in order to find the needed velocity triangles that will satisfy both the efficiency models and the Euler equation for each operating point. The Euler equation for hydraulic turbines, representing the hydrodynamic work exchanged between the fluid and the runner blades, is given by:

$$gH_n\eta_h = u_1c_{1u} - u_2c_{2u} \tag{1}$$

where indices 1, 2 denote the location at the inlet and outlet of the runner respectively, g – gravitational constant, H_n – net head, η_h - hydraulic efficiency, u – circumferential velocity at specific location, c_u – projection of the absolute velocity in the circumferential direction at specific location.

For a given rotational speed of the runner – ω and for given turbine discharge – Q, the components of the outlet velocity triangle can be calculated (see figure 2). The circumferential velocity c_{2u}, needed for the second term on the right hand side of equation (1), is calculated from the velocity triangle for infinite number of runner blades corrected by the approximated slip factor Δ_2 [18]:

$$|\overrightarrow{c_{2u}}| = \sqrt{c_2^2 - c_{2m}^2}; \quad \overrightarrow{c_2} = \overrightarrow{u_2} + \overrightarrow{w_{2\infty}} + \overrightarrow{\Delta_2}; \quad |\overrightarrow{\Delta_2}| \approx \frac{\pi u_2 \sin\beta_2}{Z_2}; \quad c_{2m} = \frac{4Q}{\phi\pi D_2^2} \tag{2}$$

where c_{2m} – meridional component of the absolute velocity c, w_∞ - theoretical relative velocity in the rotating frame of reference for infinite number of blades, ϕ – blockage coefficient at the outlet of the runner due to the thickness and angle of the runner blades, usually in the range of 0.9-0.85 [18, 19]. In the present work, the upper limit value is used for the F99 while the lower limit value is used for the RPT.

At the inlet, calculation becomes a bit more complicated due to the fact that efficiency is unknown. The available hydraulic energy of the fluid ($E\eta_h = gH_n\eta_h$) will produce total circulation of $\Gamma_1 = \oint_l c_{1u}r_1 d\varphi = 2\pi r_1 c_{1u}$ at the inlet of the runner, which will, also from the Euler equation, result in:

$$c_{1u} = \frac{\Gamma_1}{2\pi r_1} = \frac{gH_n\eta_h + u_2c_{2u}}{u_1} \tag{3}$$

where $r_1 = D_1/2$. Again, due to the relative whirl that exists between the finite number of blades at the inlet of the runner, this velocity component will be corrected by the approximate slip factor calculated as:

$$|\overrightarrow{\Delta_1}| \approx \frac{\pi u_1 \sin \beta_1}{Z_1}; \qquad \overrightarrow{c_1} = \overrightarrow{u_1} + \overrightarrow{w_{1\infty}} + \overrightarrow{\Delta_1}; \tag{4}$$

Hence, even for the BEP point, this will result in a misalignment, between the blade angle at the runner inlet and the fluid flow angle, which can be reduced by increasing the number of runner blades, by reduction of circumferential velocity or by decreasing the blade angle.

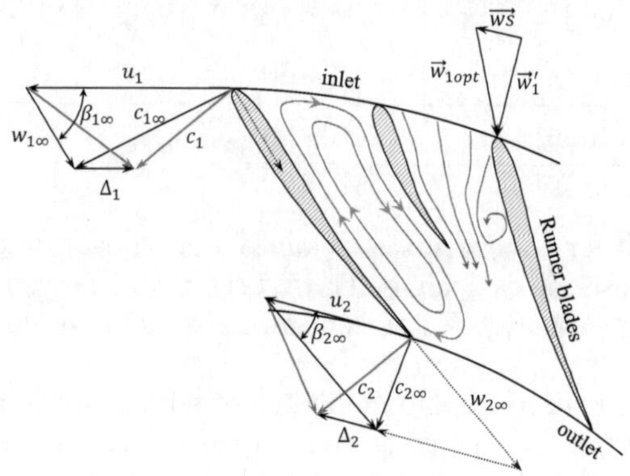

Figure 2. Illustration of velocity triangles for finite and infinite number of blades

Together with equation 3, another equation is needed for the calculation of efficiency. Simple efficiency models for the losses in the runner and draft tube are introduced hereafter.

Incidence loss at the inlet of the runner

Incidence loss is caused by the difference between the direction of the fluid flow and the fixed blade angle at the inlet of the runner (see figure 2). The fluid that is about to enter the runner will have a relative velocity $\overrightarrow{w}_1'$ that is a function of the rotational speed ω, available energy $E\eta_h$ and the discharge Q, while the fluid that has already entered the runner will quickly align to and follow the surfaces of the runner blades (i.e. relative velocity at optimum incidence $\overrightarrow{w}_{1opt}$). Assuming that the vector difference between the relative fluid velocity and the optimum incidence for the same discharge is lost ($\overrightarrow{ws} = \overrightarrow{w}_1' - \overrightarrow{w}_{1opt}$), a simplified incidence loss model can be established by writing [10]:

$$\Delta\eta_{incidence} = \frac{|\overrightarrow{ws}|^2}{2gH_n} \tag{5}$$

Swirl loss at the outlet of the runner

Swirl loss is a kinetic energy loss due to the residual whirl being created at the outlet of the runner from the mismatch of the rotational speed to the turbine discharge. The draft tube is than unable to transform this kinetic energy into pressure and this energy is lost in the lower reservoir. Depending on the type of draft tube and its stream wise increase of the draft tube cross section area, free-vortex-flow effect will reduce this kinetic energy before the fluid leaves the draft tube [1]. Nevertheless, assuming that 90% of the whirling kinetic energy at runner outlet is lost, a simplified swirl loss model can be established as:

$$\Delta\eta_{swirl} = 0.9 \cdot \frac{c_{2u}^2}{2gH_n} \tag{6}$$

Friction loss between runner blades

Friction loss in the blade channel of the runner is due to the shear forces in the boundary layers attached to the wetted surfaces. In analogy with the friction factor for calculation of pipe losses, using the average of the squared relative velocity at the inlet and outlet of the runner $\overline{w}^2 = (w_1^2 + w_2^2)/2$,

IOP Conf. Series: Journal of Physics: Conf. Series **1042** (2018) 012003 doi:10.1088/1742-6596/1042/1/012003

and the average hydraulic diameter between the inlet and the outlet of the blade channel $\bar{D}_h = (D_{h1} + D_{h2})/2$, a simplified model may be determined as:

$$\Delta\eta_{runner\_fric} = c_f \frac{L_b}{\bar{D}_h} \frac{\bar{w}^2}{2gH_n} \tag{7}$$

where c_f – skin friction coefficient. For the analysis in this paper, a constant value of the skin friction coefficient was assumed for both runners $c_f = 0.015$.

Friction and diffuser losses in the draft tube

Due to skin friction, secondary flow in the elbow and decelerating flow in the draft tube, pressure recovery will be done with some losses. For best elbow draft tubes installed in vertical turbines, Raabe [1] suggests using of the following model:

$$\Delta\eta_{DT} = 0.12 \frac{c_{2m}^2}{2gH_n} \tag{8}$$

Finally, combining all aforementioned losses yields the simplified loss model in the runner that is also used to model the total hydraulic efficiency in the Euler equation. This will introduce error in the results because total efficiency should include distributor losses, leakage losses, mixing losses, blade loading losses, disk friction losses, recirculation losses etc. However, for operation close to the BEP and in case of only examining the general shape of the hill chart, this simplified analysis should give a first clue on the expected performance of the runner and make an educated guess in case of a thorough optimization of the turbine.

$$\eta_h \approx 1 - \Delta\eta_{incidence} - \Delta\eta_{swirl} - \Delta\eta_{runner\_fric} - \Delta\eta_{DT} \tag{9}$$

The solution of the system of two equations (3) and (9) will give the values of circumferential component of the absolute velocity at the inlet of the runner c_{1u}, the hydraulic efficiency η_h and the angle of the fluid stream before entering the runner $\alpha_{1,in}$ (this angle should not be confused with the guide-vane opening angle α_{GV}), for each combination of the rotational speed of the runner – ω and the turbine discharge – Q. Net head H_n is assumed to be constant in the full range of the hill chart. Substitution of equation (9) into (3) will result into a transcendental equation that contains trigonometric functions of the dependent variable, requiring a special mathematical treatment that makes it difficult or most often impossible to solve analytically. Hence, an iterative procedure was employed to approximate the solution for each combination of ω and Q. The iteration procedure turned out to be extremely stable and, in order to achieve a rather remarkable convergence criteria of 10^{-9}, it requires only 20-40 iterations at each operating point and less than 5 seconds to compute the full hill chart, making it perfect for fast full-factorial optimization of the global parameters of the turbine [10].

4. Results and discussion

The calculation method described in section 3 was used to calculate a rather broad range of operating conditions for the two runners described in section 2. Side by side comparison of the measured and calculated hill chart for the F99 runner is given in figure 4, while the same is done for the RPT runner in figures 5 and 6. Regarding the general shape and the trend of the hill charts in figure 3, similarities can be observed with a close agreement for the position of the BEP. Also, for the theoretical calculation, the characteristic curves of zero incidence and zero swirl were precisely located and plotted on the right hand side hill chart. The zero swirl curve has a relatively low curvature for both runners and it passes through the origin of the $Q_{ED} - n_{ED}$ coordinate system. The slope of the zero incidence and zero swirl curves at the intersection point is governed by the geometry of the runner blades at the inlet and outlet respectively. Moving away from those two curves in the hill chart will increase the respective losses, and this effect, combined with the friction and draft tube losses at each point in the hill chart will yield the general shape of the efficiency isocurves. Single losses predicted for the F99 runner are presented on figure 3 for the four main losses modeled and used in equation (9). The arrows represent the direction of loss increase.

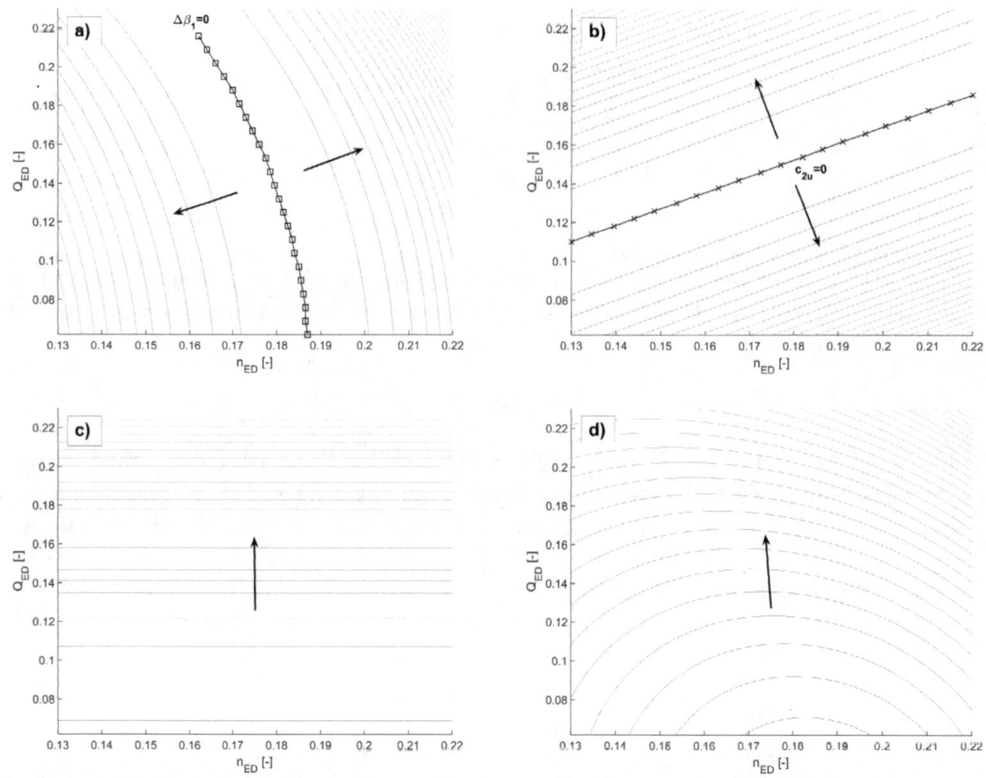

Figure 3. Efficiency decomposition into the four main losses for the F99 runner, with the arrow showing the direction of loss increase. a) contours of incidence losses, b) contours of swirl losses, c) contours of draft tube losses, d) contours of runner friction losses.

Comparing the curves of constant guide vane openings (measured) and constant flow angle at the inlet of the runner for the F99 (predicted), a significant difference can be seen for their locations in the hill chart area, with similar negative slope along the curves. This can be explained by (1) the expected nonlinear relation $\alpha_{GV}(\alpha_{1,i})$ and, (2) the losses in the circular cascade of guide vanes were not included in the analysis. If this was included in the calculation, despite the correction of the constant guide vane opening lines, prediction of absolute values of efficiency will also improve, especially in the zones away from the BEP.

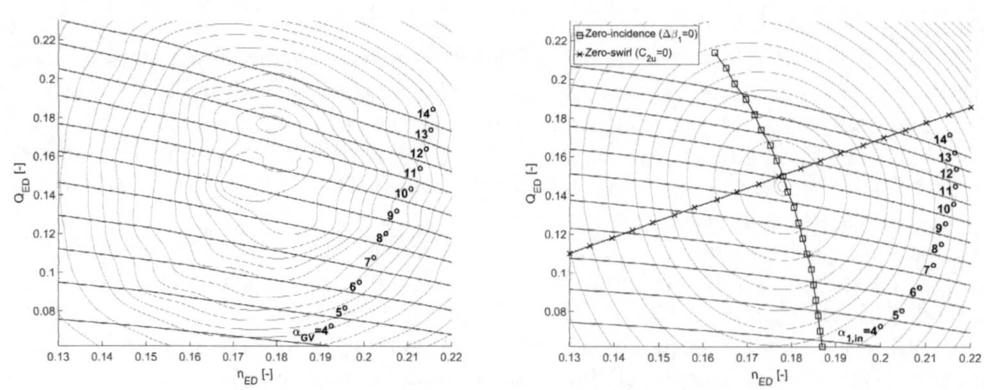

Figure 4. Hill charts of efficiency for the F99 runner.
Left side – measurements, right side – theoretical calculation

When comparing the result for the RPT runner shown on figure 5, a larger difference between measurements and theoretical calculation are observed. Comparable agreement for the general shape of the hill charts can be observed only on the right half of both hill charts (for n_{ED} larger than 0.22), while for the left half there is almost no agreement at all. The efficiency isocurves of the RPT are stretched in one direction that has a positive slope with reference to the n_{ED} axis. After careful examination of the possible reasons for the observed difference and comparing the measured data with data from previous measurements on the same models, it was concluded that the RPT runner is unexpectedly more sensitive to the Reynolds number than the F99 runner. Namely, the model test data used for this paper was obtained for open loop configuration of the test rig and at head $H_n \approx 12\ [m]$, resulting in Reynolds numbers ($Re = u_2 D_2 / \nu$) lower than $2 \cdot 10^6$ for the left half of the graph, and lower than $3 \cdot 10^6$ for the entire range. This is well below the recommended value of $Re_{min} \geq 4 \cdot 10^6$ in the IEC standard [17].

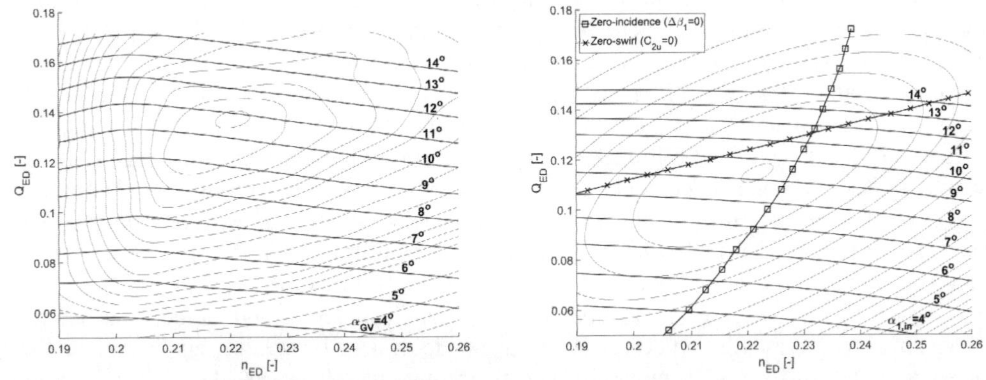

Figure 5. Hill charts of efficiency for the RPT runner.
Left side – measurements, right side – theoretical calculation

Efficiency measurements in a closed loop on the RPT model and for heads $H_n > 20\ [m]$ were previously reported by Olimstad [15]. These results are presented in figure 6, where the improvement in the Reynolds number effect is clearly visible for the left half of the hill chart previously mentioned. Additionally, the general shapes of the hill charts are more comparable to the theoretical calculation than previously seen in figure 5.

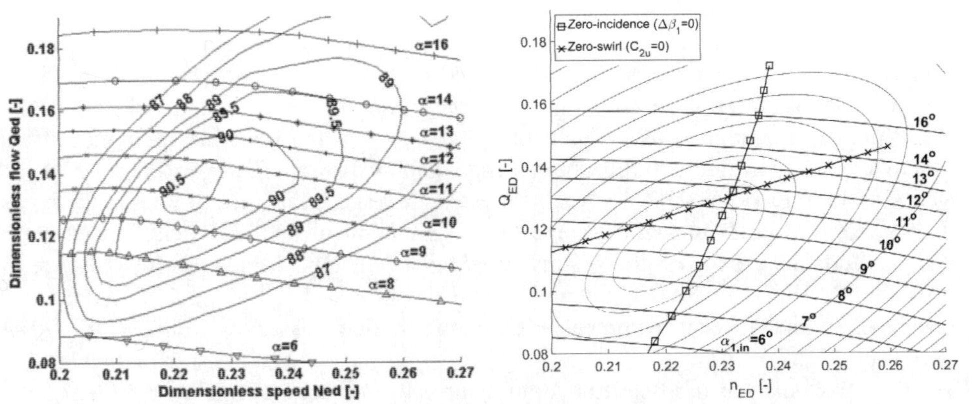

Figure 6. Hill charts of efficiency for the RPT runner.
Left side – measurements [15], right side – theoretical calculation

In the theoretical calculation, the position of the BEP is predicted to be at a lower Q_{ED} value when compared to the experimental results, while the n_{ED} value is pretty close in both cases. Similarly as for the F99 runner, there is an expected difference in the position of the curves of constant α_{GV} and $\alpha_{1,in}$ with a similar negative slope along the curves.

5. Conclusions

The theoretical analysis for the general shape of the hill chart presented here gives promising results using minimal geometric information for the turbine as the input. The turbine models selected for the analysis represent a rather special and interesting case since the only difference is the blade geometry of the runner, which nevertheless, gave large differences in the measured hill charts. The models used for predicting the losses in the runner and draft tube are extremely simple and intuitive and yet this was enough to give a comparable prediction for the hill chart shape of both runners and have provided a better understanding on which parameters are responsible for it.

The geometry of the shroud profile of the blade was used to simplify the complex three dimensional geometry of the runner where velocity triangles at the inlet and outlet of the runner were sought to satisfy the system of the Euler equation and the efficiency model. The numerical method used is also simple and stable, providing calculation of the full hill chart in less than 5 seconds. Based on the presented calculation and side-by-side comparison with experimental results, it was demonstrated that the shape of the hill charts is greatly dependent and driven by the geometry at the inlet and the outlet of the runner, namely the channel width, runner diameter and blade angle at the inlet and runner diameter and blade angle at the outlet. This is also characterized by the position and shape of the zero-incidence and zero-swirl curves in the hill charts, representing the joint hydrodynamic contribution from the inlet and outlet geometry of the runner.

The observed difference between absolute values of efficiency and guide vane opening at different operating conditions is expected and strongly dependent on the sophistication of the loss models used. Results can always be improved if more accurate models are used and more loss sources are simultaneously analyzed, such as the losses in the guide and stay vanes, blade loading losses, recirculation losses, disk friction and clearance gap leakages, etc. For prediction of the constant guide vanes opening curves, geometric relations and loss model is also needed. Using the free vortex flow and the radius of the trailing edge for the guide vanes, calculated values of $\alpha_{1,in}$ can be extrapolated and used as an input for such model.

6. References

[1] Joachim R 1985 Hydropower – the design, use, and function of hydromechanical, hydraulic, and electrical equipment *VDI Verlag*

[2] Grigori I K 1994 Hydraulic machines: turbines and pumps (Translated from Russian) *Lewis Publishers*

[3] Emil F M 1991 Water power development *Akademiai Kiado*

[4] Xin L, Yongyao L, Bryan W K, Weizheng W 2015 A selected literature review of efficiency improvements in hydraulic turbines *Renewable and Sustainable Energy Reviews* **51** 18-28

[5] Jurgen S, Helmut B, Helmut J 2016 Analysis of the Leakage Behaviour of Francis Turbines and Its Impact on the Hydraulic Efficiency - A Validation of an Analytical Model Based on Computational Fluid Dynamics Results *ASME. J. Fluids Eng.* **139(2)** 021106 (11 pages)

[6] Osterwalder J, Hippe L 2010 Guidelines for efficiency scaling process of hydraulic turbo machines with different technical roughness of flow passages *Journal of Hydraulic Research* **22:2** 77- 102

[7] Hermod B 1996 Analysis of losses in hydraulic turbines *Proceedings of the 18th IAHR symposium* 294-303

[8] Aungier R H 1995 Mean streamline aerodynamic performance analysis of centrifugal compressors *ASME J. of Turbomachinery* **117** 360-366

[9] Oh H W, Kim K Y 2001 Conceptual design optimization of mixed-flow pump impellers using mean streamline analysis *Proc. Instn. Mech. Engrs* **215** 133-138
[10] Pei-Yuan L, Chu-Wei G, Yin S 2015 A new optimization method for centrifugal compressors based on 1D calculation and analysis *Energies* **8** 4317-4334
[11] Galvas M R 1972 Analytical correlation of centrifugal compressor design geometry for maximum efficiency with specific speed *NASA TN* ***D-6729*** 1-42
[12] Risberg S, Jonassen M, Jonassen R 2008 Design of Francis turbine runners based on a surrogate model approach *The International Journal of Hydropower and Dams* **15(5)** 80-84
[13] Demeulenaere A, Ligout A, Hirsch C 2004 Application of multipoint optimization to the design of turbomachinery blades *Proc. of ASME Expo: Power for Land, Sea and Air* **5** 1481-1489
[14] Semenova A, Chirkov D, Lyutov A, Cherny S, Skorospelov V, Pylev I 2014 Multi-objective shape optimization of runner blade for Kaplan turbine *IOP Conf. Ser.: Earth Environ. Sci.* **22** 012025
[15] Grunde O, Torbjorn K N 2012 Stability limits of reversible-pump turbines in turbine mode of operation and measurements of unstable characteristics *ASME Fluids Eng.* **134** 111202
[16] Chirag T, Michel J C, Gandhi B K, Ole G D 2013 Experimental and numerical studies for a high head Francis turbine at several operating points *ASME Fluids Eng.* **135(11)** 111102
[17] IEC-60193, Hydraulic Turbines, Storage Pumps and Pump-turbines: Model Acceptance Tests, International standard, *International Electro-technical Commission*, 3, CH-1211 Geneva, Switzerland, 16 November, p. 578, 1999
[18] Geroff V 1973 Water turbines (In Bulgarian) *D. I. „Tehnika" Sofia*
[19] Stepanoff A J 1948 Centrifugal and axial flow pumps. Theory, design and application. *John Willey & Sons, Inc.*

IOP Conf. Series: Journal of Physics: Conf. Series **1042** (2018) 012004 doi:10.1088/1742-6596/1042/1/012004

Numerical investigation on performance and sediment erosion of Francis runner with different guide vane profiles

R Lama[1], D R Dahal[1], S Gautam[1], N Acharya[1], H Neopane[1*] and B S Thapa[1]
[1]Department of Mechanical Engineering, Kathmandu University, Dhulikhel, Nepal
*Corresponding author (hari@ku.edu.np)

Abstract. Francis turbine are ideal turbines for Himalayan and Andes region where both low and high-altitude mountains are located. Turbines operating in such regions face operational and maintenance problems due to the sediment erosion. In order to reduce the erosion effects on these components the design of components for higher sediment handling is essence. This paper presents performance analysis of Francis runner and prediction of sediment erosion on the runner blades for different operating conditions with different guide vane profiles. The simulations were carried out for 11 guide vane opening angles using Tabakoff erosion model. At full load and best efficiency point the erosion was localized at pressure side of runner blades outlet due to higher relative velocity. On the other hand, at part load condition, erosion was observed at suction side of the blades. Application of asymmetric guide vane profile NACA 4412 showed higher efficiency for all operating conditions with minimum erosion on runner blades in compare to symmetric guide vane profile NACA 0012.

Keywords: Francis Turbine, Sediment Erosion, Leakage Flow, Vortex, Efficiency.

1. Introduction

World's energy consumption is rapidly increasing due to economic growth and structural changes in the countries. As countries develop and living standards improve, energy demand grows rapidly. Among other sources of energy: Hydropower is the leading renewable source for electricity generation globally, supplying 71% of all renewable electricity. The global hydropower capacity increased by more than 30% between 2007 and 2015 accounting to a total of 1209 GW in 2015 [1]. Thus, hydropower has become most popular, efficient and reliable sources of renewable energy supplying 77% of all renewable energy resources [2].

Hydro-energy plays an important role in Nepal's economic future because of the scale of its potential. It is estimated that Nepal has the technically and economically viable potential to put in place over 43,000 megawatts (MW) of hydro generation capacity, however, the installed capacity has reached about 800 MW [3]. If such potential is realized, increasing energy demand of Nepal i.e. 7–9% per year could easily meet [2]. However, the political instability, weak policy of the government, topography and geological conditions of the nation are key constraints for the development of Hydropower in Nepal. Nepal has huge potential as well as technical challenges for hydropower development in Himalayan region due to sedimentation and erosion problems. The tropical climate, immature geology, and intense seasonal rainfall are the main reasons for this problem. As a result, river in this region transport substantial amount of sediments during the monsoon [4]. Francis turbines are widely used in many hydropower projects due to its efficiency and flexibility. It converts both the pressure and kinetic energy retained in water to mechanical energy inside runner thereby producing Electrical energy when coupled to the generator. Francis turbines are ideal turbines for the Himalaya and Andes areas where both the high and low altitude mountains are located. Past research has shown that up to about 80% of hydro turbines required for future installations under Himalayan basins in Nepal will need Francis type of turbine [5].

2. Current status and literature study

The erosion of turbine components due to effect of sediments in the water is one of the major operational challenges in development of hydropower in Himalayan and Andes region. The conventional design criteria for high head Francis turbine i.e. The highest possible efficiency avoiding cavitation with smallest possible dimension is not appropriate for projects under these regions, where sedimentation is severe .In Nepal, the rivers in the Himalayan region contain high percentage of hard abrasive minerals like quartz, which causes rapid erosion of turbine components and affects the performance of turbines [6]. This in turn decreases the efficiency, reliability and operating life of the turbines and creates a serious challenge in operation of the hydropower project [4].

The currently adopted methods such as the prevention of sedimentation in the catchment areas, tapping sediments at intakes, and applying preventive coatings on turbine components exposed to high velocity water are not only complete solution for reducing sediment erosion [6]. The sediment erosion problem in this region leads to create a new design philosophy which incorporate this problem in hydraulic design of turbine. The reduction of relative velocity at runner outlet is one of the alternative solutions to prevent sediment erosion in Francis turbines [7]. For better performance of turbines operated in sediment laden water guide vane profile also plays a significant effect. Proper hydraulic design of guide vane with smoothest possible acceleration, reduction of clearance gaps between guide vanes and facing plates, careful choice of stay vane outlet angle, and new material are the key factors for their better performance in sediment laden water [8].

Past studies have shown that various factors: characteristics of sand particles, angle of impingement, type of turbine materials, way of operation, and hydraulic design of turbines contribute to the performance of turbine. Among them, it is recognized that optimized design of hydraulic turbines is the key parameter for obtaining better performance thereby reducing sediment erosion problems. This study mainly focus on the prediction of erosion pattern and erosion intensity on runner blades at different operating conditions using different guide vane profiles. Also, other focus of study was performance analysis of turbine at all operating conditions for reference runner with symmetric and asymmetric guide vane profiles.

3. Numerical model

3.1 Governing Equations and Erosion Model

The numerical analysis was carried out in the exact dimension of Jhimruk hydropower plant of Nepal which consists of 24 guide vanes and 17 runner blades excluding the splitters. The domain for steady analysis consists of 4 stationary guide vanes and 3 rotating runner vane passages. The simulations were carried out for 1 symmetric guide vane profile and 2 asymmetric guide vane profiles with 11 guide vane opening angles. The mass flow rate at inlet and atmospheric pressure at outlet were chosen as boundary conditions for the simulation. The designed mass flow rate of the turbine representing 100% flow was taken as best efficiency point and mass flow rates for different guide vane opening angles were calculated based on the best efficiency point and the corresponding flow by maintaining the almost constant head for all the cases. The numerical analysis focus on the prediction of turbine performance at different guide vane opening angles with 3 different guide vane profiles and prediction of the critical eroded areas of the runner blades. For this study Shear Stress Transfer turbulence (SST) model and Tabakoff erosion model were used.

3.1.1 Governing equations

For a steady, turbulent, three-dimensional flow laws of conservation of mass (continuity) and momentum are used to describe the flow. The governing equations for the fluid flow analysis were given by Navier-Stokes equation defined in terms of conservation of mass and momentum as,

$$\frac{\partial \rho}{\partial t}+\frac{\partial}{\partial x_j}(\rho u_j)=0 \tag{1}$$

$$\rho\left(\frac{\partial u_i}{\partial t}+u_j\frac{\partial u_i}{\partial x_j}\right)=-\frac{\partial p}{\partial x_i}+\frac{\partial \tau_{ij}}{\partial x_j}+S \tag{2}$$

The analysis was performed using Reynolds Averaged Navier-Stokes (RANS) equations.

$$\frac{\partial}{\partial t}(\rho\overline{u_i})+\frac{\partial}{\partial x_j}(\rho\overline{u_i}\overline{u_j})=\frac{\partial}{\partial x_j}\left\{-\overline{p}\delta_{ij}-\rho\langle u_i'u_j'\rangle+\mu\left(\frac{\partial\overline{u_i}}{\partial x_j}+\frac{\partial\overline{u_j}}{\partial x_i}\right)\right\} \tag{3}$$

Mainly two types of turbulence model are used for analysis of hydraulic machinery i.e. K-epsilon (k-ε) and Shear Stress Transport (SST). For this study Shear Stress Transport (SST) was used that accounts transport of turbulent shear stress at boundary layer and away from boundary regions for both boundary flow and separated flows [9] [10] . The two-equation model is given as;

$$\frac{\partial(\rho k)}{\partial t}+\frac{\partial(\rho u_j k)}{\partial x_j}=P-\beta^*\rho\omega k+\frac{\partial}{\partial x_j}\left[(\mu+\sigma_k\mu_t)\frac{\partial k}{\partial x_j}\right] \tag{4}$$

$$\frac{\partial(\rho\omega)}{\partial t}+\frac{\partial(\rho u_j\omega)}{\partial x_j}=\frac{\gamma}{v_t}P-\beta\rho\omega^2+\frac{\partial}{\partial x_j}\left[(\mu+\sigma_k\mu_t)\frac{\partial\omega}{\partial x_j}\right]+2(1-F_1)\frac{\rho\sigma_{\omega 2}}{\omega}\frac{\partial k}{\partial x_j}\frac{\partial\omega}{\partial x_j} \tag{5}$$

Where, Pressure, turbulent eddy viscosity and turbulent shear stress is computed from:

$$P=\tau_{ij}\frac{\partial u_i}{\partial x_j}$$

$$\tau_{ij}=\mu_t\left(2S_{ij}-\frac{2}{3}\frac{\partial u_k}{\partial x_k}\delta_{ij}\right)-\frac{2}{3}\rho k\delta_{ij}$$

$$S_{ij}=\frac{1}{2}\left(\frac{\partial u_i}{\partial x_j}+\frac{\partial u_j}{\partial x_i}\right)$$

$$\mu_t=\frac{\rho a_1 k}{\max(a_1\omega,\Omega F_2)}$$

3.1.2 Erosion Model

Sediment in waters flowing in rivers of Nepal contains large contents of quartz which has hardness greater than turbine material. For numerical analysis of sediment erosion, quartz having density 2650 kg/m^3 and size 100 μm with particle number 1000 was used at the inlet of domain. Mass flow rate of sediment was taken as 0.07 kg/s. The erosion rate of sediment can be calculated using two different models which are available in Ansys CFX i.e. Finnie's erosion model and Tabakoff's erosion model.

Finnie Erosion model

The abrasion erosion rate (ER), i.e. ratio of mass of material removed to the mass of particle in the carrier fluid is given by,

$$ER=\frac{P_n\rho_m}{S_w} \tag{6}$$

This ER depends upon the kinetic energy of impinging particle and incidence angle α:

$$ER=KV_P^n f(\alpha) \tag{7}$$

With,

$$f(\alpha)=\frac{1}{3}\sin^2\alpha \qquad 0<\alpha<0.4\pi \tag{8}$$

$$f(\alpha)=\sin 2\alpha-3\cos^2\alpha \qquad 0.4\pi<\alpha<0.5\pi \tag{9}$$

Tabakoff Erosion model
The Erosion Rate (ER) is given by,

$$E=k_1 f(\gamma)V_P^2\cos^2\gamma[1-R_T^2]+f(V_{PN}) \tag{10}$$

where,

$$f(\gamma)=[1+k_2k_{12}\sin(\gamma\frac{\pi/2}{\gamma_0})]^2$$

$$R_T=1-k_4V_P\sin\gamma$$

$$f(V_{PN})=k_3(V_P\sin\gamma)^4$$

$$k_2=\begin{cases}1.0 & if\,\gamma\le 2\gamma_0\\ 0.0 & if\,\gamma>2\gamma_0\end{cases}$$

In this study, Tabakoff's erosion model was adopted as it considers more parameters and is more reliable than Finnie's model. [11]

4. Mesh sensitivity analysis

The passage domain for this study consist of 4 GVs and 3 runner blades, composed of 0.7 million hexahedral cells created in ICEM CFD. The boundary condition applied for the case was mass flow rate inlet with 2350 kg/s for whole domain at BEP and atmospheric pressure outlet. For mesh sensitivity analysis NACA 0012 profile and GV opening at BEP was selected. Guide vane hub and shroud tips were added at stator domain to be 1mm.
Numerical uncertainty was evaluated using Grid Convergence Index (GCI) taking reference from procedure for estimation and reporting of discretization error in CFD application by Ismail B. Celik [12]. For uncertainty analysis, three mesh were created by increasing number of elements in each mesh with global size factor of 1.3. Thus, number of elements in each mesh were 0.1 million, 0.3 million and 0.8 million. Efficiency of the turbine with three different meshes were observed and uncertainty in the efficiency was measured.
Table 1 shows uncertainty in measurement of efficiency of the turbine that is calculated to be 2.85% for fine mesh. Hence fine mesh with 0.8 million elements were selected for overall numerical simulation of the research which is shown in figure 1.

Table 1. Discretization in error in numerical simulation

Efficiency as a variable measured for coarse, medium and fine mesh	
N_1, N_2 & N_3	0.8M, 0.3M & 0.1M
r_{21}	1.39
r_{21}	1.36
ϕ_1	92.67
ϕ_2	92.95
ϕ_3	92.66
P	0.396
ϕ_{ext}^{21}(extrapolated)	95.07
e_a^{21}	0.003162
e_{ext}^{21}	0.002232
GCI_{fine}^{21}	**2.85%**

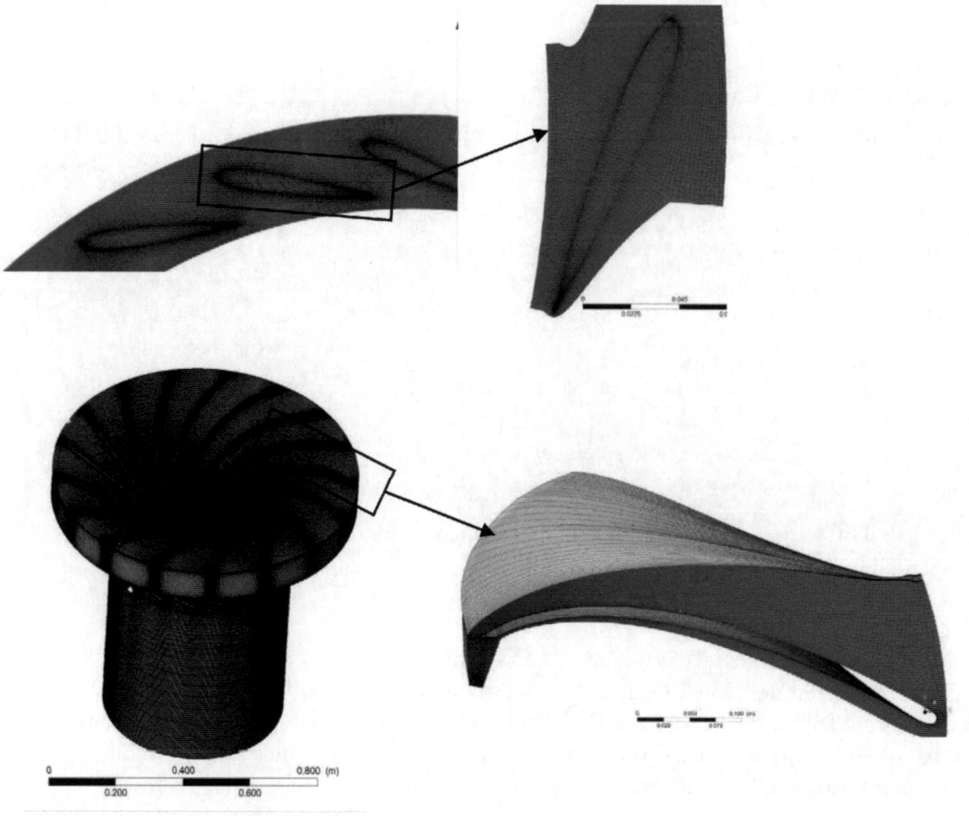

Figure 1. Fine Mesh for Guide vane and Runner

5. Results and Discussion

5.1 Efficiency comparison for different guide vane profiles

Figure 2 shows comparison of efficiencies of turbine corresponding to different guide vane profiles for different operating conditions with sediments. Similarly, Table 2 represents GV opening angle with respect to BEP and its corresponding mass flow rates. Flow percentage is represented by guide vane opening angles where, 100% flow resembles the best efficiency point.The 0 degree guide vane opening angle implies the 100% flow which is equal to the designed mass flow rate of the turbine i.e. 2350 kg/s. Similarly, -5 degree represents 70.23% flow and 5 degree represents 129.76 % flow. At part load conditions, due to the minimum flow the efficiencies corresponding to all the three NACA profiles is lowest.The efficincy curves for all the 3 guide vane profiles showed similar trends for all operating conditions. It was observed that among efficiencies of the runner corresponding to the 3 different guide vane profiles , the runner containing NACA 4412 guide vane profile showed the optimum effficiency, whereas runner associted with NACA 0012 guide vane profile showed minimum efficiency.On the other hand, runner having NACA 2412 guide vane profile showed the intermediate efficienciy lying in between the NACA 0012 and NACA 4412. When using the asymmetric guide vane profle NACA 4412 in place of NACA 0012, the efficiency was increased in the range of 2-3.5%, where as only 0.5-1.2% efficincy was increased in case of NACA 2412. The rise in efficiency in case of asymmetric guide vane profile was due to the less leakage flow in asymmetric guide vane profile.

Table 2. Guide Vane opening angle and corresponding mass flow rate with respect to BEP

Gv angle (degree)	-5	-4	-3	-2	-1	0	1	2	3	4	5
Flow (kg/s)	1650.6	1790.48	1930.36	2070.24	2210.12	2350	2489.88	2629.76	2769.64	2909.652	3049.9
%	70.23	76.19	82.14	88.09	94.04	100	105.95	111.90	117.85	123.8	129.76

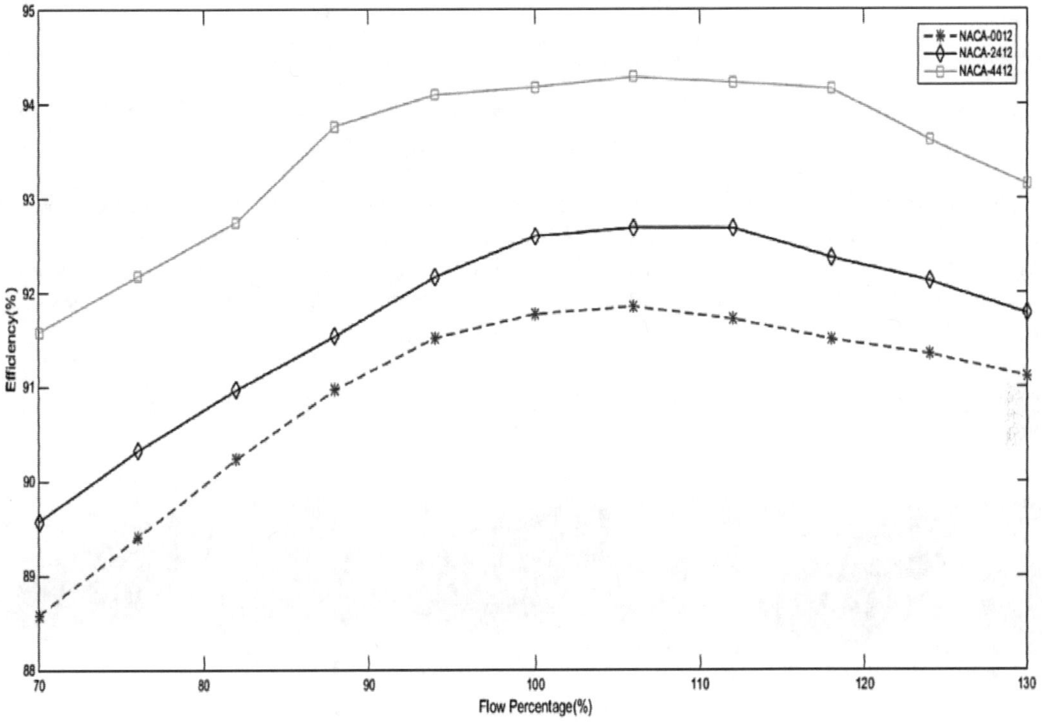

Figure 2. Efficiency of the runner for all operating conditions for 3 GV profiles

5.2 Erosion on runner blades for different operating conditions with 3 guide vane profiles

The effects of 3 guide vane profiles on the erosion of runner blades was the main focus of this study. Figure 3 to 5 show the erosion pattern and its severity on the runner blades at pressure side with different hydrofoils and operating conditions. The erosion rate density was used as sediment measuring parameter for analysis. At design load and full load condition the erosion was concentrated on the middle and outlet areas of the blades but the intensity of the erosion was higher at full load condition which was due to the higher relative velocity at runner outlet as a result of higher flow rate in case of full load condition. Due to the minimum flow in case of part load condition i.e. assuming minimum sediment inflow rates compare to the design load and full load, the erosion observed were less on the runner blades since erosion is directly proportional to the sediment inflow rates. The erosion on runner blade with symmetric guide vane profile NACA 0012 showed higher erosion on the runner blades, whereas at asymmetric guide vane NACA 2412 and NACA 4412 showed lesser erosion on runner blades at all operating conditions. Significant erosion on runner blades at suction side of the runner blades were observed at part load condition whereas at design load and full load, the erosion on suction side of the runner blades were negligible.

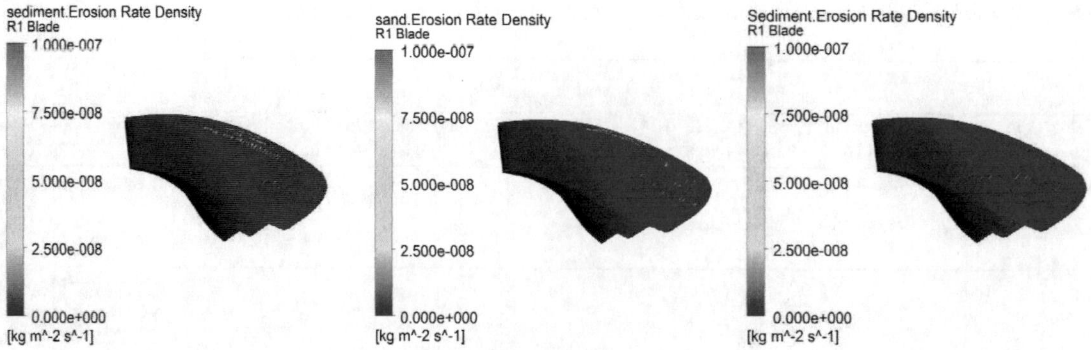

NACA 0012 NACA 2412 NACA 4412

Figure 3. Erosion on runner blades at best efficiency point

From observation of erosion on runner blades with NACA 0012 at part load, erosion at inlet areas are more sever but in case of NACA 2412 and NACA 4412, the middle and outlet areas were more vulnerable to the erosion. The NACA 4412 having minimum pressure difference and minimum leakage flow in compare to NACA 2412 and NACA 0012 showed minimum erosion on runner blades at all operating conditions.

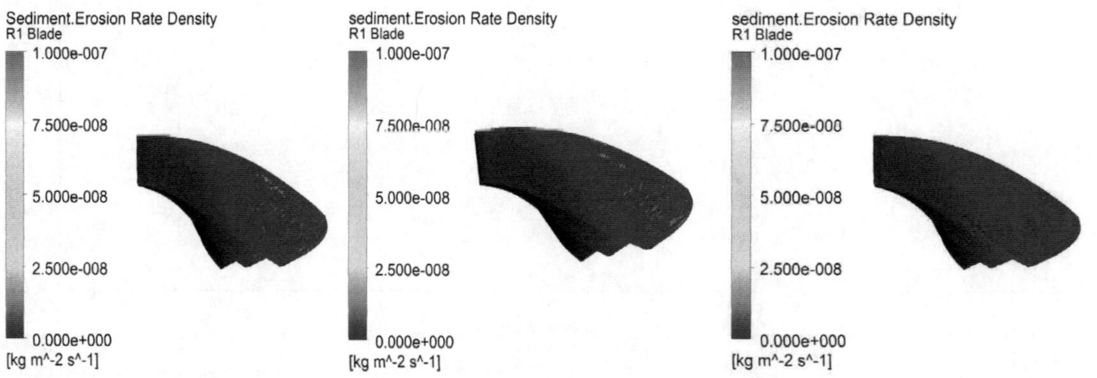

NACA 0012 NACA 2412 NACA 4412

Figure 4. Erosion on pressure side of runner blades at full load condition

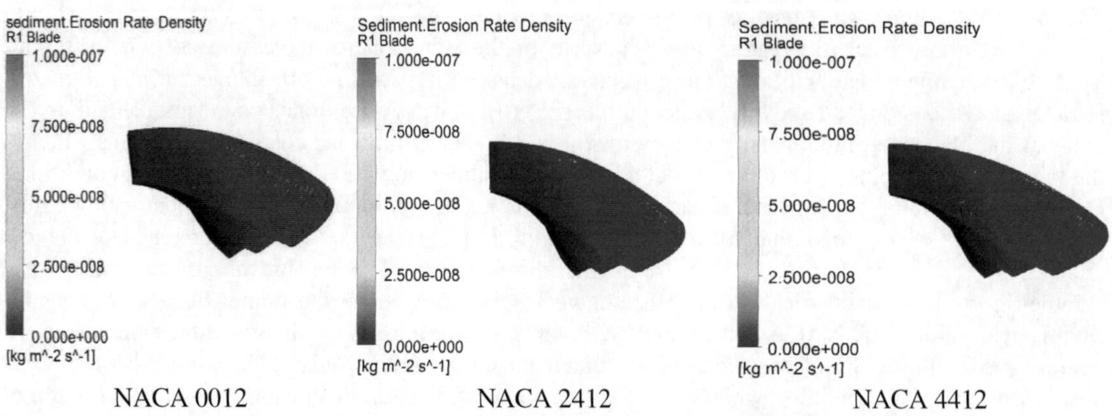

NACA 0012 NACA 2412 NACA 4412

Figure 5. Erosion on pressure side of runner blades at part load condition

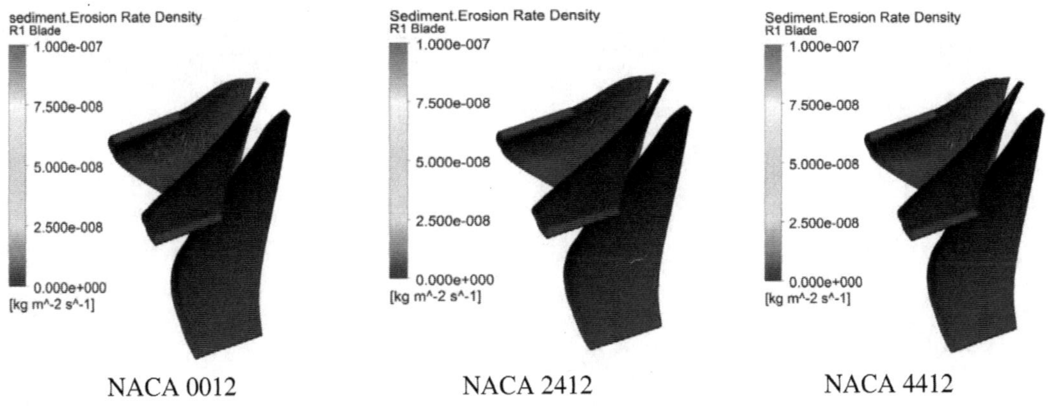

Figure 6. Erosion on pressure side of runner blades at suction side at part load condition

Erosion at suction side of the blade was also observed while running the simulation at part load condition. The stagnation angle at part load condition is different from the ideal stagnation angle as in best efficiency loading. Due to variance in stagnation angle from ideal case, turbulence was observed at suction side of the blade which was the main cause of erosion on this side of the blade.

5.3 Leakage flow pattern in NACA 4412

Figure 7 shows the nature of vortices in NACA 4412 originated form leakage flow. At full load condition the vortex has tendency to hit the adjacent GV profiles rather than passing towards runner inlet due to negative leakage flow. However, the generated vortex follows its path from GV trailing edge to runner inlet at part load and best efficiency point. The negative leakage flow in NACA 2412 and NACA 0012 were found to be negligible in comparison to NACA 4412.

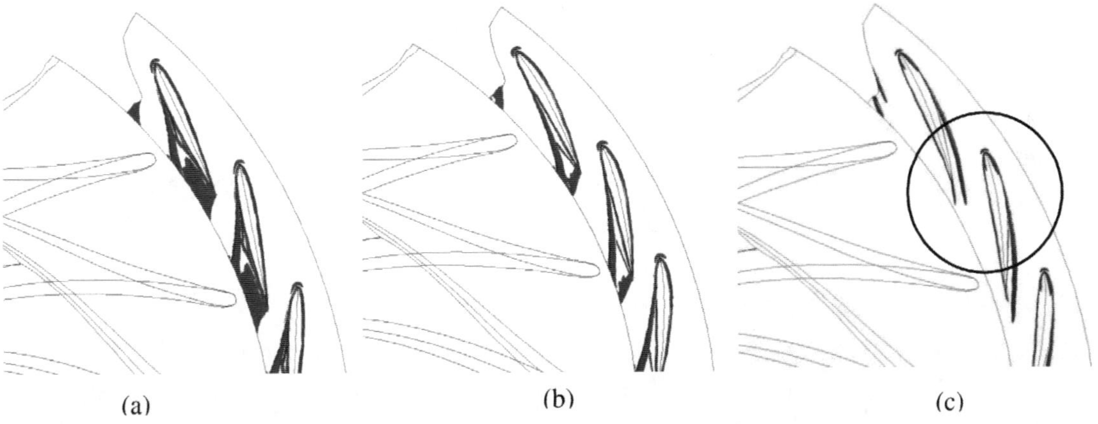

Figure 7. Leakage flow patterns in all operating conditions at NACA 4412 at (a) part load (b) best efficiency (c) full load

Behaviour of particle flow around the runner blade loaded at BEP for 3 GV profiles is shown in figure 8. Laminar streamlines on runner blade were observed in case of NACA 4412 due to minimum leakage flow while turbulent streamlines were observed in case NACA 2412 and NACA 0012 as the leakage flow increase from NACA 4412 to NACA 0012. Due to the reduction of size of the vortex in NACA 4412 in comparison with NACA 0012, the simultaneous effects of secondary flow was seen minimum in the NACA4412 hydrofoils as shown in figures 8.

IOP Conf. Series: Journal of Physics: Conf. Series **1042** (2018) 012004 doi:10.1088/1742-6596/1042/1/012004

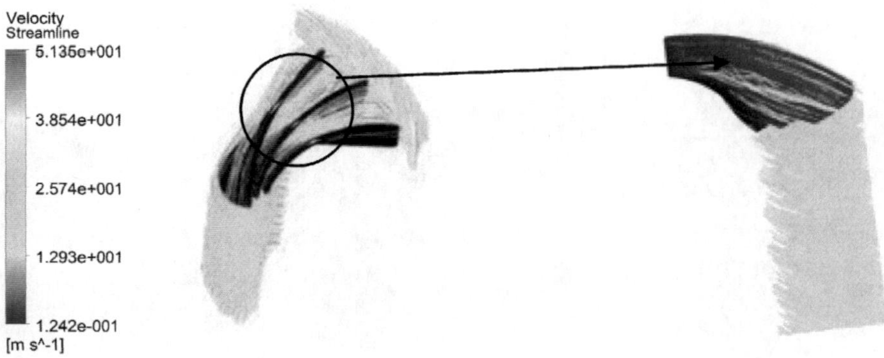

(a)

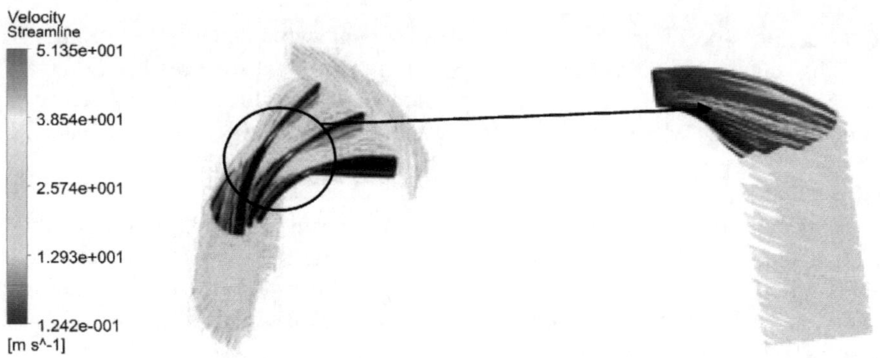

(b)

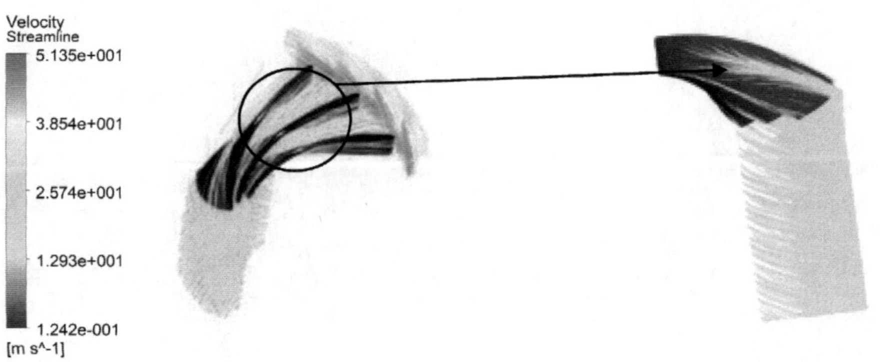

(c)

Figure 8. Particle flow around the runner blade at best efficiency point for 3 GV profiles (a) NACA 0012 (b) NACA 2412 (c) NACA 4412

6. Conclusion

Numerical analysis on Francis turbine of Jhimruk Hydropower Plant was carried out with three different guide vane profiles with sediments. Performance, pattern of erosion and its intensity on runner blades were observed for different operating conditions. At design load and full load condition erosion pattern was almost similar and dominant to the pressure side of the blade. Nevertheless, at full load condition intensity of erosion was higher due to increased mass flow rate. At part load condition high turbulence was created because stagnation angle shifts significantly from idealness which leads to flow separation causing erosion on suction side of the runner blades. The effects of symmetric and asymmetric GV profiles on performance of turbine were studied. Study showed rise in efficiency for NACA 4412 and NACA 2412 in reference to NACA 0012 by 2-3.5% and 0.5 -1.2 % respectively for all operating ranges. Out of three guide vanes profiles NACA 4412 exhibited optimum performance and minimum sediment erosion on runner blades for all operating conditions. Implementation of NACA 4412 in real power plant can be done to have minimum sediment erosion on runner blades as well as guide vanes and optimum performance at all operating conditions. However, experimental verification is needed to verify the observed results.

References

[1] World Energy Council *2016 World Energy Resources* (London : World Energy Council)
[2] Investment Board 2017 *Nepal Energy Sector Profile (*Kathmandu : Pentagram)
[3] Nepal Electricity Authority 2016 *A Year in Review- Fiscal Year 2015/2016* (Kathmandu: Nepal Electricity Authority)
[4] Neopane H P 2010 *Sediment Erosion in Hydro Turbines* PhD [dissertation] Norwegian University of Science and Technology (2010).
[5] Pantha S, Lamsal M, Thapa B S and Thapa B 2014 Prediction of turbine needed for future hydropower in nepal *Journal of Water Energy and Environment* **14** (2014) 23-26.
[6] Thapa B 2004 *Sediment Erosion in Hydraulic Machinery* PhD [dissertation] Norwegian University of Science and Technology.
[7] Thapa B S, Thapa B and Dahlhaung O G 2012 Current research in hydraulic turbines for handling sediments *Energy* **47** pp 62-69.
[8] Brekke H 2001 *Design of Hydraulic Machinery Working in Sand Laden Water in Abrasive Erosion and Corrosion of Hydraulic Machinery*, ed C G Duan and V Y Karelin (London: Imperial College Press) pp 155-181.
[9] Koirala R, Neopane H P, Shrestha O, Zhu B and Thapa B 2017 Selection of guide vane profile for erosion handling in Francis turbines *Renewable Energy* **112** (2017) pp 328-336.
[10] Chitrakar S, Thapa B S, Dahlhaug O L,Neopane H P 2017 Numerical and experimental study of the leakage flow in guide vanes with different hydrofoils *Journal of Computational Design and Engineering* **4** (2017) pp 218-230.
[11] Kang M W, Park N, Suh S H 2016 Numerical study on sediment erosioin of Francis turbine with different operating conditions and sediment inflow rates *Procedia Engineering* **157** (2016) pp 457-464.
[12] Ismail B C 2008 Procedure for estimation and reporting of uncertainty due to discretization in CFD applications *J. Fluids Eng* **130** (7).

Acknowledgments

This study was a part of PhD degree at Kathmandu University. The program is funded by EnergizeNepal. Authors would like to thank Turbine Testing Lab, Sailesh Chitrakar, Aatmaram Kayastha for their continuous help and support.

IOP Conf. Series: Journal of Physics: Conf. Series **1042** (2018) 012005 doi:10.1088/1742-6596/1042/1/012005

Pressure Pulsation in a High Head Francis Turbine Operating at Variable Speed

D. B. Sannes[1*], I. Iliev[1], E. Agnalt[1], O. G. Dahlhaug[1]
[1]Department of Energy and Process Engineering, NTNU, Norway
*Corresponding author (danielbsannes@gmail.no)

Abstract. This paper presents the preliminary work of the master thesis of the author, written at the Norwegian University of Science and Technology. Today, many Francis turbines experience formations of cracks in the runner due to pressure pulsations. This can eventually cause failure. One way to reduce this effect is to change the operation point of the turbine, by utilizing variable speed technology. This work presents the results from measurements of the Francis turbine at the Waterpower Laboratory at NTNU. Measurements of pressure pulsations and efficiency were done for the whole operating range of a high head Francis model turbine. The results will be presented in a similar diagram as the Hill Chart, but instead of constant efficiency curves there will be curves of constant peak-peak values. This way, it is possible to find an optimal operation point for the same power production, were the pressure pulsations are at its lowest. Six points were chosen for further analysis to instigate the effect of changing the speed by ±50 rpm. The analysis shows best results for operation below BEP when the speed was reduced. The change in speed also introduced the possibility to have other frequencies in the system. It is therefore important avoid runner speeds that can cause resonance in the system.

Keyword: Francis Turbine, Variable Speed, Pressure Pulsation, Hill Chart, Hydro Power

1. Introduction

The electrical grid is a complex system, which is coupled with different types of energy sources as hydro, wind and solar. Because of the flexibility that the Hydropower turbines offers, it is used to stabilize the grid by changing operating point, and therefore the power output [3]. The result of this is that the turbine more often is running off-design. Off-design phenomena such as power swings, rotating vortex rope and part load pressure pulsations can drastically reduce a runners operating life [5]. One of the problems with Francis turbines today is the formation cracks in the runner due to pressure fluctuations, and the industry if facing problems with both new and old Francis turbines [2]. It is therefore of interest to find methods for reducing the effects of pressure fluctuations, and prolong the lifetime of the runner. One potential method of doing this, is by introducing variable speed technology. This enables us to change operation point by changing the speed of the runner to where pressure pulsations are at its lowest.

The present work aims to investigate the pressure pulsations for the whole operating range, and to understand how pressure pulsation varies with different operational conditions. A total of 269 different operational points have been measured, from guide vane opening of 1 degree to 14 degree (full opening). By analysing these different operational conditions, it is possible to find out where to operate the turbine were the pressure pulsations are at its lowest. Six operational points, Full Load

(FL), High Load (HL), Best Efficiency Point (BEP), Part Load (PL), Minimum Load (ML) and Speed-No-Load (SNL) has been chosen for further analysis to check how the peak-peak values and frequency spectrum of pressure pulsation varies when the speed is changed by ±50 rpm.

1.1. Variable Speed Operation

A Francis turbine is designed for fixed speed and for one best efficiency point. The speed of the turbine is predetermined by the combination of pole pairs in the generator and the grid frequency. Therefore, turbine power output is controlled by changing the guide vane opening and thereby water discharge. By utilizing a frequency converter, together with the synchronous generator, it is possible to operate with variable speed. The generator is then decoupled from the grid, and free to operate at different speeds [5]. The turbine will be more flexible, when it can regulate both speed and water discharge, and can avoid operational conditions with cavitation problems and high pressure pulsation. The drawback is additional converter losses that must be considered.

2. Laboratory Setup and Methods

2.1. Francis Model Test Rig

Measurement were conducted at the Francis model test rig located at the Waterpower laboratory at NTNU. The test rig was operated by using an open loop configuration, as seen in Figure 1. The water is pumped from the basement reservoir to a channel located at the top of the laboratory. The water is then going through an upstream pressure tank before entering the turbine. Downstream the draft tube, there is a pressure tank which is used to regulate the pressure at the outlet from the runner. For this measurement the water level in the draft tube pressure tank was open to air. Water is then going back to the reservoir. The open loop configuration gives a uniform water level (≈12m) and provides same conditions for all operation points. The Francis turbine is a scaled down model (1:5.1) of the Tokke turbine. It has 14 stay vanes, 28 guide vanes and 30 runner blades (15 full length and 15 splitter blades).

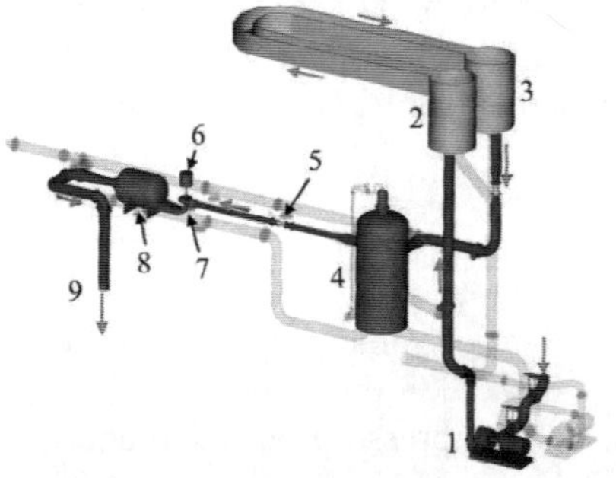

Figure 1. Open loop configuration. (1) centrifugal pump, (2) and (3) open water channel, (4) upstream pressure tank (5) flowmeter, (6) generator, (7) Francis turbine, (8) downstream pressure tank, (9) water outlet.

2.2. Logging Program and Equipment

The measurement system is a complete setup with all the necessary equipment for calculating the efficiency (and to draw a Hill Chart), according to the IEC60193. There is also additional pressure sensors along the guide vanes, on the draft tube cone and on-board the runner, according to Figure 2 and Figure 3. The draft tube pressure sensors (PTDT5, PTDT7, PTDT13, PTDT17) are mounted on the draft tube cone. Pressure sensors along guide vanes (PTGV1-PTGV6) is mounted on the top of the casing, and the on-board pressure sensors (PTR1-PTR6) is mounted between the blades on the crown. The on-board system is transmitting measurement signals via slip rings to the logging system. The logging frequency was 10240Hz for all sensors. A LabVIEW program, developed in the laboratory, was used as logging software.

2.3. Measurement Procedure

Measurements were done for guide vane opening from 1 degree to 14 degree opening (1, 2, ..., 14). The starting point for each guide vane opening was N_{ED}=0.08, and the end point was run-away speed (≈0.28). The increment between each measurement was N_{ED}=0.01. This gave a total number of 269 measurement points, from deep part load to full load. All operational points were logged for 60 seconds.

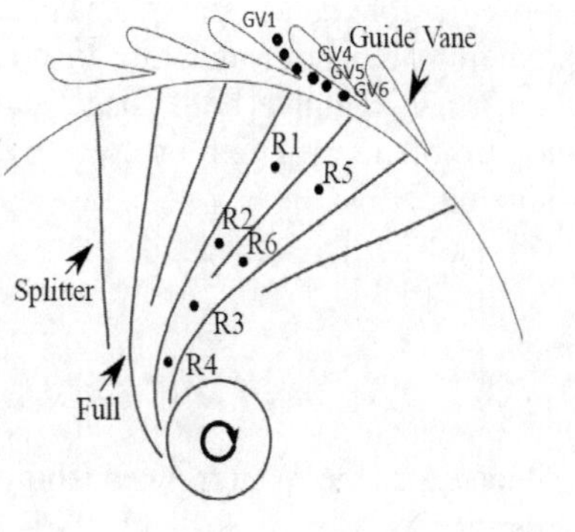

Figure 2. Pressure sensors. GV1-GV6 along guide vanes and R1-R6 on-board the runner.

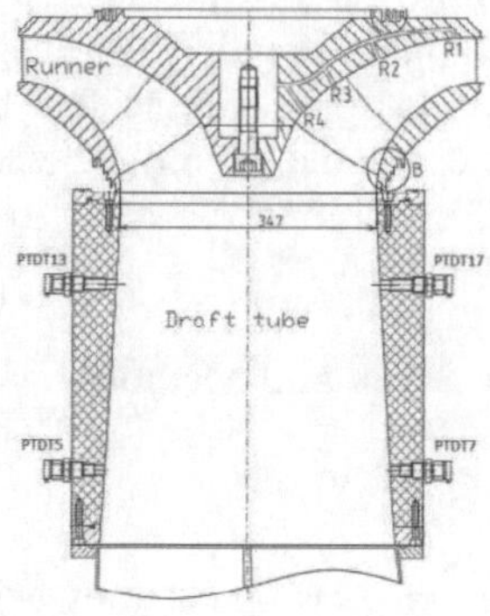

Figure 3. On-board sensors, R1-R4, and draft tube pressure sensors; DT5, DT7, DT13, DT17.

2.4. Data Analysis

MatLab were used for analysing all the measured data. Efficiency calculations for the Hill Chart was done according to IEC60193 [6]. For analysing peak-peak values of pressure data, the Histogram method was used, with a confidence level of 97% as suggested in IEC60193 [6]. The Histogram method has proven to give good results, and have been used by Gogstad in his PhD [8] and been suggested by Dörfler et.al. [7]. For the spectral analysis, Welch method was used together with a Hanning window with 50% overlap. A set of filters were applied to the pressure data to remove noise, before the data was used for analysis.

3. Results

Figure 4 shows the complete Hill Chart for the Francis turbine, from guide vane opening of 1 degree to 14 degree. The turbine would normally operate at N_{ED}≈0.18 (333rpm), and BEP is located at Q_{ED}≈0.154 with and maximum efficiency just above 92%. For the pressure pulsation analysis three different pressure sensors were used to obtain peak-peak values, PTDT13, PTR2 and PTGV4 (Figure 2 and Figure 3). These three sensors were chosen to investigate pressure pulsation effects in the whole turbine based on their location. Peak-peak values was calculated for all operation points, for each sensor, and are presented in a similar way as the Hill Chart. Instead of lines of constant efficiency, there is lines of constant peak-peak values. By doing this, it is easier to identify were pressure pulsation have high values, and which operational areas that should be avoided. In these diagrams, the 97% confidence level is used, as suggested in [6]. Other values may be more correct, and 99% was used by Gogstad [8], but this was not investigated further in this work. Figure 5 shows the results from the draft tube sensor, PTDT13. Diagrams for the other sensors can be found in the Appendix. Peak-peak values are normalised, based on the BEP value. The absolute values are given in Table 2.

Six points were chosen for further analysis; SNL, ML, PL, BEP, HL and FL. Table 1 shows more detailed information on the different operational point. Each point was compared with another operational point, with the same power output, were the speed is reduced and increased by 50 rpm, to investigate how the pressure pulsation varies with a change in speed. The different points can be seen as dots in Figure 5 (along N_{ED}=0.155, N_{ED}=0.1795 and N_{ED}=0.208). The lines that goes through the different points are lines of constant power. Table 2-4 shows how peak-peak values varies when the speed was changed.

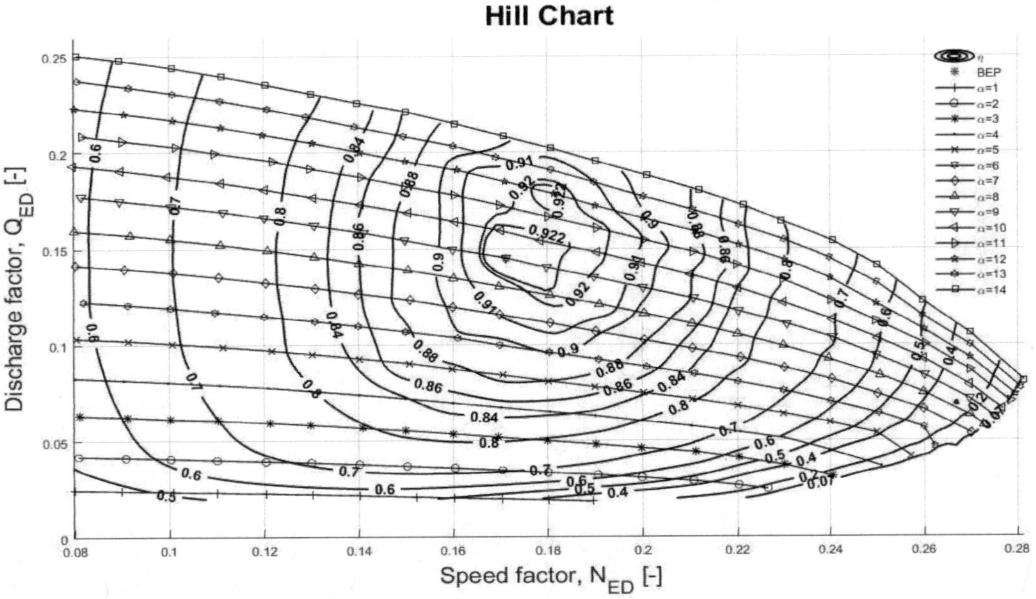

Figure 4. Hill Chart for the whole operation range. Guide vane opening from 1 degree to 14 degrees. N_{ED} varies from 0.08 to run away speed.

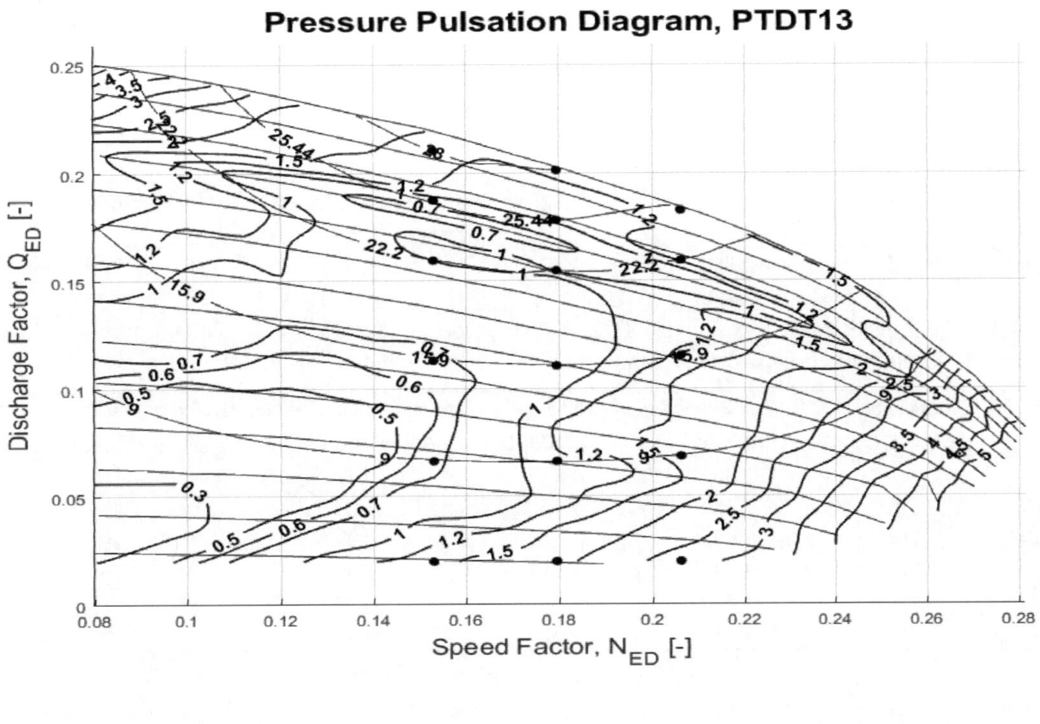

IOP Conf. Series: Journal of Physics: Conf. Series **1042** (2018) 012005 doi:10.1088/1742-6596/1042/1/012005

Figure 5. Pressure Pulsations diagram for PTDT13, draft tube sensor. Peak-peak values are normalised based on BEP value and are calculated based on 97% confidence level. The BEP value is calculated to be 2.39kPa. The marked points is used for further analysis, and the lines that goes through them is lines of constant power.

Table 1. Detailed information for operational points used in further analysis

	GV opening. [degree]	Efficiency [%]	N_{ED} [-]	Q_{ED} [-]	Speed [rpm]	Mech Power [kW]
ML	4	84.9	0.1798	0.067	338	9.04
PL	7	91.3	0.1795	0.111	336	15.9
BEP	10	92.2	0.1795	0.155	335	22.2
HL	12	92.2	0.1795	0.178	334	25.4
FL	14	90.7	0.1795	0.201	333	28.0
SNL	1	40	0.1795	0.02	340	1.3

Table 2. Peak-peak values for sensor PTDT13, 97% confidence level. The last two columns show the relative change in peak-peak between normal speed and changed speed. All values are in kPa.

	Normal speed (≈333rpm)	Reduced speed (≈286rpm)	Increased speed (≈386rpm)	Relative change when reducing speed	Relative change when increasing speed
ML	2.82	1.62	3.86	0.57	1.37
PL	2.16	1.62	3.04	0.75	1.41
BEP	2.39	2.39	2.95	1.0	1.23
HL	2.52	2.13	3.29	0.85	1.31
FL	3.12	3.1	-	0.99	-
SNL	1.88	1.31	2.74	0.70	1.46

Table 3. Peak-peak values for sensor PTGV4, 97% confidence level. The last two columns show the relative change in peak-peak between normal speed and changed speed. All values are in kPa.

	Normal speed (≈333rpm)	Reduced speed (≈286rpm)	Increased speed (≈386rpm)	Relative change when reducing speed	Relative change when increasing speed
ML	2.96	2.56	3.73	0.86	1.26
PL	2.87	2.7	3.51	0.94	1.22
BEP	2.37	2.76	2.59	1.16	1.09
HL	2.5	2.54	4.22	1.02	1.69
FL	3.1	2.82	-	0.91	-
SNL	1.15	0.78	1.59	0.68	1.38

Table 4. Peak-peak values for sensor PTR2, 97% confidence level. The last two columns show the relative change in peak-peak between normal speed and changed speed. All values are in kPa.

	Normal speed (≈333rpm)	Reduced speed (≈286rpm)	Increased speed (≈386rpm)	Relative change when reducing speed	Relative change when increasing speed
ML	4.1	2.86	4.84	0.70	1.18
PL	2.86	2.95	4.05	1.03	1.42
BEP	3.04	3.02	3.91	0.99	1.29
HL	3.05	3.32	4.44	1.09	1.46
FL	3.35	3.29	-	0.98	-
SNL	4.82	3.19	5.91	0.66	1.22

IOP Conf. Series: Journal of Physics: Conf. Series **1042** (2018) 012005 doi:10.1088/1742-6596/1042/1/012005

Frequency spectrum analysis was conducted for the six different points and the ones with changed speed. This provides information on which flow phenomena that causes pressure fluctuation in the turbine, and which is contributing the most. By using reduced frequencies, it is possible to compare the results when the speed is changed, and see if there are changes in the occurring frequencies. Pressure data was analysed with Welch method in MatLab. Figure 6 shows the result from spectrum analysis for sensor PTGV4 at High Load operation. Figure 7 shows the results for PTR2 at High Load operation.

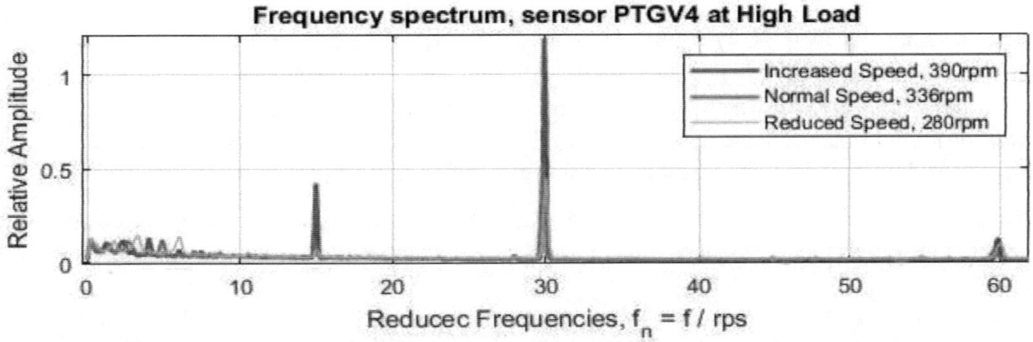

Figure 6. Frequency analysis of pressure data from PTGV4 at HL, for normal speed and changed speed. The blade passing frequency is most dominant with a frequency of 30 times the rotational frequency (f). There is also an effect due to the splitter blade that causes frequencies at 15*f. Stochastic pulsations occur in the low frequency region.

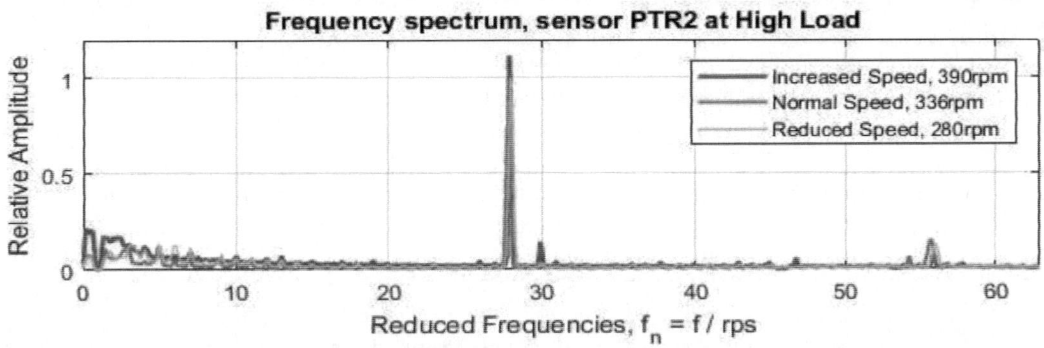

Figure 7. Frequency analysis of pressure data from PTR2 at HL, for normal speed and changed speed. The guide-vane-passing frequency is most dominating (28*f). There is also stochastic pulsations in the low frequency region.

4. Discussion

The three different pressure pulsation diagrams show different behaviour for the pressure pulsations. The draft tube pressure pulsation clearly shows a trend towards lower peak-peak values for lower speed and discharge, while the pulsation around the guide vanes and on-board the runner shows lowest values at BEP and around an area where the speed is reduced. There are also indications that it is possible to find optimum operation points for minimal pulsations, especially when reducing the speed, but there will be a give-and-take between lowest peak-peak values in the different part of the turbine.

The diagrams from PTGV4 and PTR2 show more similarities than the one for PTDT13. This make sense because Rotor-Stator Interaction (RSI) is the most dominating phenomena for both sensors (Figure 6 and Figure 7). It is therefore important to figure out what is most damaging for the turbine, and find optimal operation based on this.

From Table 2-4 we can see the result from changing the speed (±50 rpm) from six operation points. There was no effect at BEP for the three sensors. This is as expected since this is the design point. Also, when the speed is increased, the pressure pulsations are increasing for all operation points. On the other hand, if the speed was reduced, pressure pulsations dropped for almost all the sensor at all operation points, and the reduction is highest for operation below BEP. In the draft tube, the pressure pulsations are reduced with 25 %, 30% and 30% for PL, ML and SNL. While on-board the runner (PTR2), the same reduction is 0%, 29% and 34%, and 6%, 14% and 32% for PTGV4. Pressure pulsations for operation above BEP was not strongly affected by the reduction in speed, and peak-peak values on-board the runner increased with 9% at HL. It should be said that speed change is not limited to ±50 rpm, (15% change in speed), and different speed may give even lower pressure pulsations.

Frequency analysis was performed for all operation points, to investigate which pulsation phenomena that occurred for the different points, and to find out how the change in speed affected the frequencies. Figure 6 and Figure 7 shows the results from HL for PTGV4 and PTR2. For these two sensors, the RSI was found to be most dominant for all the operation point, except at part load operation, where the draft tube vortex rope was dominating. Also the effect of the splitter blades was found in the analysis, which was half the blade-passing frequency, along the guide vanes. For PTDT13, the pressure pulsations was dominated of a frequency between 1.45-1.8 times the runner frequency, combined with stochastic fluctuations. At part-load operation, the vortex rope was dominating the pressure pulsation. This proves some of the challenges that follows variable speed operation. Many of the frequencies are speed depended, such as RSI and the draft tube vortex rope. A change in speed will therefore introduce many more frequencies that potentially can cause resonance in the system.

5. Conclusion

An experimental investigation of a high head Francis turbine was conducted to investigate pressure pulsation for the whole operation range. The purpose of this was to see it is possible to reduce the effect of pressure pulsation by utilizing variable speed, and thereby avoid operational problems and crack formation in the runner. A Hill Chart was constructed, and pressure pulsation were analysed for three different sensors. Three pressure pulsation diagrams were created, similar to a Hill Chart but with lines of constant pressure pulsations (peak-peak values). Six points were chosen for further analysis to see how the pressure pulsation varied with a speed change of ±50 rpm, but with the same power output.

Increasing the speed resulted in higher peak-peak values for off-design operation, while reducing the speed resulted in reduced pressure pulsation in most of the cases. The highest reduction in pressure pulsation was found for operation below BEP, for all three sensors. The maximum reduction was found for Speed-No-Load, were the peak-peak values was reduced by more than 30%. The investigated points were chosen arbitrarily, and the pressure pulsation diagrams gives indication that other operational points would give even lower peak-peak values. RSI was found to be most dominating in the upper part of the turbine, together with the vortex rope at curtain part load operation points. These pressure fluctuations are speed dependent and will change according to the speed. It is there important avoid runner speed that can cause resonance in the system.

Appendices

A.1 PTGV4 Pressure Pulsation Diagram

Peak-peak values are calculated with 97% confidence level. All the values are normalise with respect to the BEP value. The BEP value is calculated to be 2.37kPa. Normal operating is along N_{ED}=0.18.

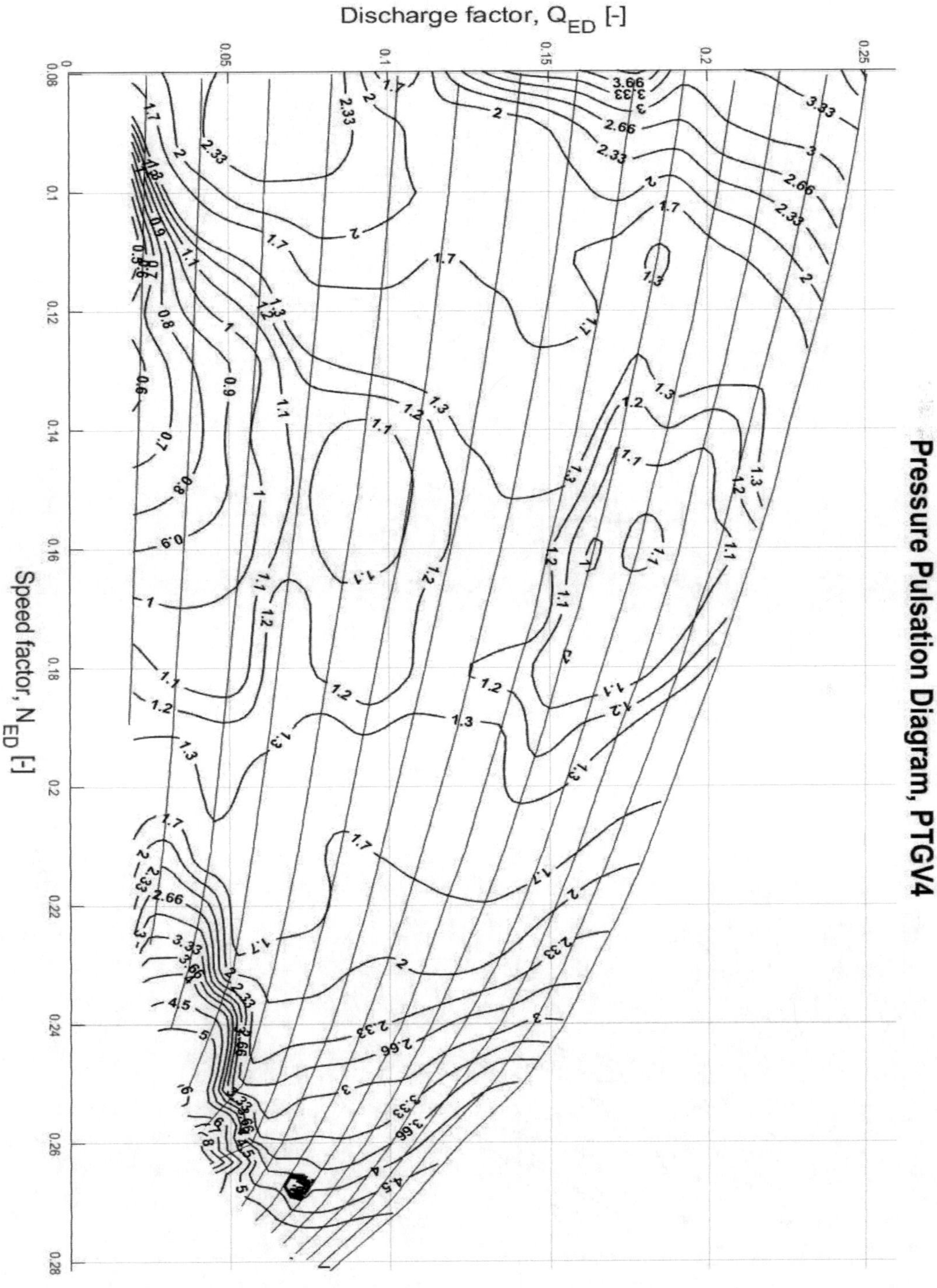

IOP Conf. Series: Journal of Physics: Conf. Series **1042** (2018) 012005 doi:10.1088/1742-6596/1042/1/012005

A.2 PTR2 Pressure Pulsation Diagram

Peak-peak values are calculated with 97% confidence level. All the values are normalise with respect to the BEP value. The BEP value is calculated to be 3.04kPa. Normal operating is along N_{ED}=0.18.

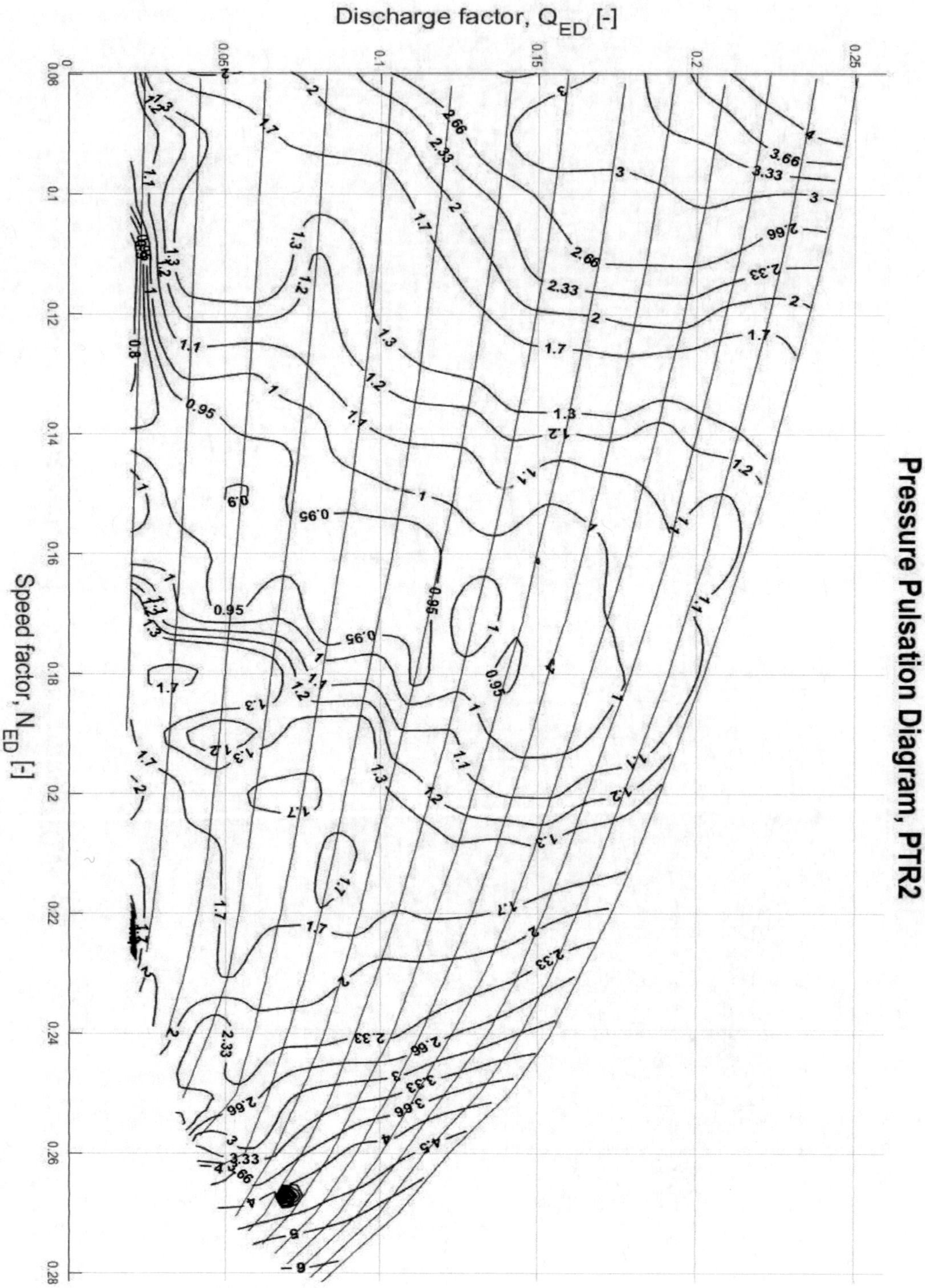

References

[1] Trivedi C, Agnalt E, Dahlhaug O G, 2017, *Investigations of unsteady pressure loading in Francis turbine during variable-speed operation,* Elsevier, Renewable Energy 113 (2017) 397-410.

[2] Norwegian Hydropower centre, *High Head Francis – HiFrancis,* https://www.ntnu.edu/nvks/hifrancis, read; 25.02.2018

[3] Trivedi C, Cervantes M J, Gandhi B K, Dahlhaug O G, 2013, *Experimental and Numerical Studies for a High Head Francis Turbine at several Operating Points,* ASME, Vol. 135.

[4] Bergan C, Dahlhaug O G, 2016, *Experimental Investigation of a High Head Model Francis Turbine During Steady-State operation at Off-Design Conditions*, CRHT2016-18, KU.

[5] Valavi M, Nysveen A, 2016, *Variable-Speed Operation of Hydropower Plants: Past, Present, and Future,* XXII International Conference on Electrical Machines (ICEM), 2016, pp 640-646.

[6] IEC 60193:1999, *Hydraulic turbines, storage pumps and pump-turbines – Model acceptance tests,* (1999-11-16).

[7] Dörfler P, Sick M, Coutu A, 2013, *Flow-Induced Pulsation and Vibration in Hydroelectric Machinery.* Springer London

[8] Gogstad P J, 2017, *Experimental investigation and mitigation of pressure pulsations in Francis turbines*, Doctoral Thesis at NTNU, 2017:202.

IOP Conf. Series: Journal of Physics: Conf. Series **1042** (2018) 012006 doi:10.1088/1742-6596/1042/1/012006

Study on the effect of the runner design parameters on 50 MW Francis hydro turbine model performance

Ujjwal Shrestha[1], Zhenmu Chen[1] and Young-Do Choi[2*]

[1]*Graduate School, Department of Mechanical Engineering, Mokpo National University, Youngsan-Ro, Cheonggye-Myeon, Muan-Gun, Jeonnam, 58554, South Korea*
[2]*Department of Mechanical Engineering, Institute of New and Renewable Energy Technology Research, Mokpo National University, Youngsan-Ro, Cheonggye-Myeon, Muan-Gun, Jeonnam, 58554, South Korea*
* *Corresponding author (ydchoi@mokpo.ac.kr)*

Abstract

Francis hydro turbine is the dominant turbine in the hydropower generation. Francis turbine has been installed at most 60% of the hydropower in the world at present. Although the basic design for the Francis turbine has various method regarding the specific speed. The runner meridional shape varies with different specific speed. Despite having, the basic design but there is still some room for the optimization. In this study 50 MW, Francis hydro turbine with specific speed 323 m-kW was designed and considered for the optimization. The various parameter as runner meridional shape (curve profile of hub, shroud, leading edge and trailing edge), blade angle and its distribution, blade thickness, runner inlet width that has been considered for the optimization of the runner for enhancement of the performance.

Keywords: Francis hydro Turbine, Design, Meridional Shape, Blade Angle.

1. Introduction

Francis hydro turbine has a wide range of operating condition from low head to high head. There have been lots of study on the high head Francis runner but unfortunately, there is not sufficient study on low head Francis runner. Due to lack of study on low head Francis turbine the exploitation of low head potential energy is limited and mainly depend on the axial turbine. The low head Francis runner has different characteristics than the high head one. The difference between low head and high head Francis runner is the dependency of the flow discharge with rotational speed, which is inverse in high head Francis Runner, and direct with low head Francis runner [1]. Francis runner design is influenced by numerous parameter due to the shape complexity and nature of the three-dimensional flow. The major parameter for the design of Francis runner is rotational speed of the turbine, guide vane height, meridional shape of the runner, runner inlet and outlet diameter, blade angle, blade angle distribution from leading edge to trailing edge, thickness distribution profile [2]. Every power plant has its own specification for turbine design to operate at maximum efficiency. In this article, the effect of runner design parameters on the performance of the turbine has been discussed. The specification of the 50 MW Francis turbine is shown in the table 1.

Table 1: Parameter of 50 MW Francis Turbine.

Parameter	Value
Head	38.7 m
Discharge	132 m^3/s
Power	46.3 MW
Rotational Speed	150 rpm
Specific Speed	323 m-kW

IOP Conf. Series: Journal of Physics: Conf. Series **1042** (2018) 012006 doi:10.1088/1742-6596/1042/1/012006

2. Turbine Blade Design Parameter

Many runner design parameters affect the turbine performance. Runner meridional shape, blade angle distribution, and thickness distribution in developing the blade shape has a major contribution to increase or decrease the performance of the turbine. Based on non-dimensional specific speed, Francis turbine design parameter is determined. P Henry and M Hussain Khan [3] [4] have used the equation 1 for determination of non-dimensional specific speed. The parameters for the meridional shape are the function of the non-dimensional specific speed.

$$\upsilon = \frac{\omega\sqrt{Q/\pi}}{(2.E)^{3/4}} \tag{1}$$

2.1. Meridional Shape

The meridional shape is the main concern with blade shape and design. The meridional shape has an influence on the flow characteristics of runner. The meridional shape of the runner is varied according to the specific speed of the runner. The meridional shape is generated from the equation developed by BOVET [3]. The meridional shape can adjust according to the need of the turbine shape. The curve for the meridional shape generally belongs to above curve family represented by equation 2.

$$\frac{y}{y_{max}} = \frac{16}{3\sqrt{3}}\sqrt{\frac{x}{l}\left[1-\frac{x}{l}\right]^3} \tag{2}$$

The characteristics curve shown in figure 1 represent the curve profile for hub and shroud with the variable with x represents the length of the curve. The maximum vertical height of the curve at the 25% of the total length of the curve. Similarly, with help of the characteristic curve, the meridional shape for the 50 MW Francis runner has been generated in Blade Gen ANSYS 18.1 [5].The l and r are the length and radius, subscript i and e represent hub and shroud profile respectively, and subscript 1 and 2 represent leading edge and trailing edge respectively. However, in the meridional shape, the shroud profile length is 25% of the total length of the curve, which is the location of maximum camber.

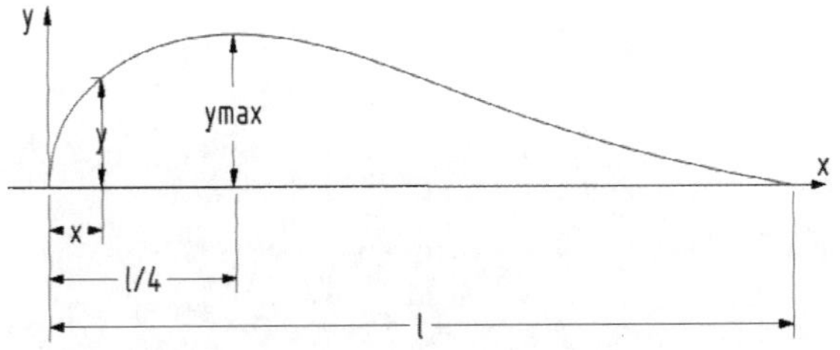

Figure 1: Characteristic curve for Flow Channel

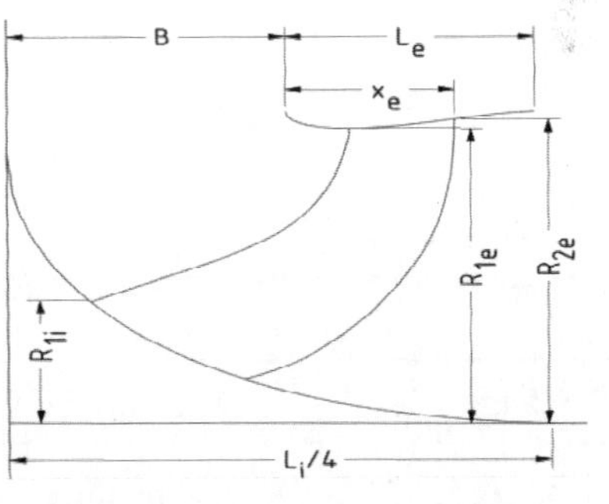

Figure 2: Nomenclature of the Meridional Shape

The empirical dimension of meridional shape as shown in figure 2 can be calculated by using equations 3 to 8 [3]. After calculating the empirical values, it is should be converted to the real physical value by using some constant and represented by the upper case in figure 2. The constant

IOP Conf. Series: Journal of Physics: Conf. Series **1042** (2018) 012006 doi:10.1088/1742-6596/1042/1/012006

value will vary according to the required runner dimension. It can determine with help of energy coefficient and discharge coefficient of the runner.

$$b = 0.8(2-\upsilon)\upsilon \tag{3}$$

$$l_i = 3.2 + 3.2(2-\upsilon)\upsilon \tag{4}$$

$$x_e = 0.52 - 0.2\upsilon \tag{5}$$

$$l_e = 2.4 - 1.9(2-\upsilon)\upsilon \tag{6}$$

$$r_{1i} = 0.4 + \frac{0.18}{\upsilon} \tag{7}$$

$$r_{2e} = \frac{1}{0.96 + 0.2\upsilon} \tag{8}$$

3. Numerical Method

The commercial CFD software ANSYS CFX 18.1 [5] has been used to perform all the analysis of the turbine. The analysis of both one pitch and the full domain has been conducted. The accuracy of the simulation is dependent on the parameters such as numerical scheme, computational grid quality, boundary condition and proper turbulence model. The casing and draft is unstructured grid consists of a tetrahedral pyramid and prism elements. Stay vane and guide vane consists of unstructured hexahedral elements from ANSYS ICEM 18.1 [5]. Hexahedral elements for the runner is generated from ANSYS TURBOGRID 18.1 [5].

Table 2: Boundary Conditions for Analysis

Analysis Type	Steady State
Turbulence Model	SST
Inlet	Total Pressure
Outlet	Static Pressure
Walls	No slip
Working Fluid	Water at 25^0C

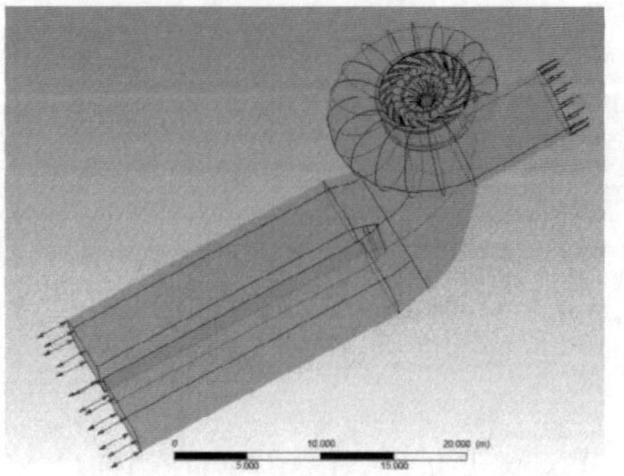

Figure 3: Full Domain Analysis

The result of the mesh independency test has been shown in figure 4. The efficiency and power output for the different mesh number has been normalized by the efficiency and power output of fine mesh. The deviation of normalized efficiency and power with respect to the variation of a mesh is less than 10%. From the mesh independency test, the minimum node number for the simulation is 450000 to reduce the result deviation less than 0.5%. Similarly, the result of turbulence model test is presented in figure 5. In turbulence model test the efficiency and power output have been normalized by efficiency and power output of SST turbulence model. The difference in the normalized efficiency and power output is less than 5%, which is compared with six different turbulence model. From the mesh independency and turbulence model test it can be verified the result of the simulation.

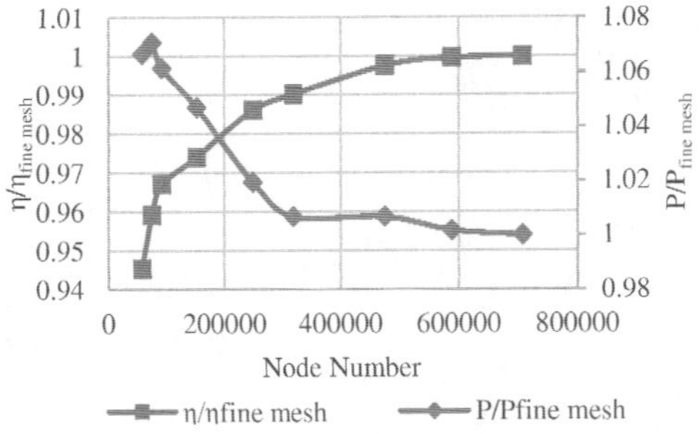

Figure 4: Mesh Independency Test

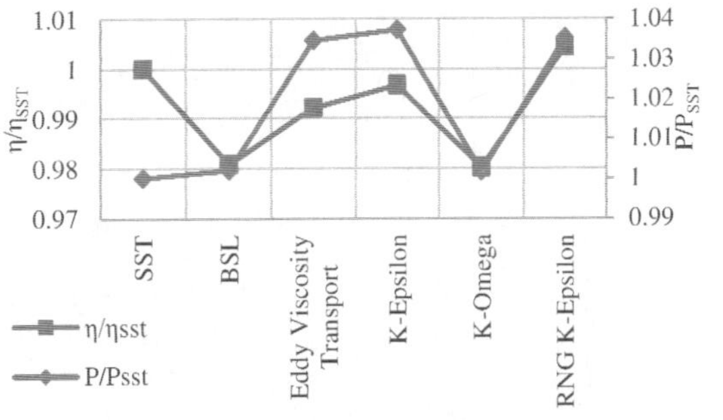

Figure 5: Turbulence Model Test

4. Result and Discussion

4.1 Blade Angle distribution

The inlet and outlet blade angle is vital for the proper runner design. The conventional way of the designing considered the inlet blade angle is fixed from the hub to shroud and outlet angle vary with the diameter. However, in the low head Francis runner, both inlet and outlet blade angle varies from hub to shroud. The inlet and outlet blade angles are calculated as per the equation 9 and 10 respectively [6]. The inlet and outlet blade angle are dependent upon the diameter of the runner. At the shroud, the inlet and outlet blade angle is nearly equal due to the minimum difference in the inlet and outlet diameter whereas in the hub due to the high difference inlet and outlet angle deviation is also high.

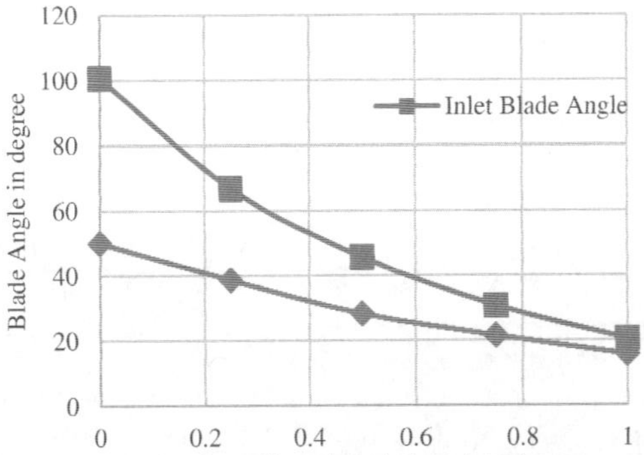

Figure 6: Blade angle variation from shroud to hub

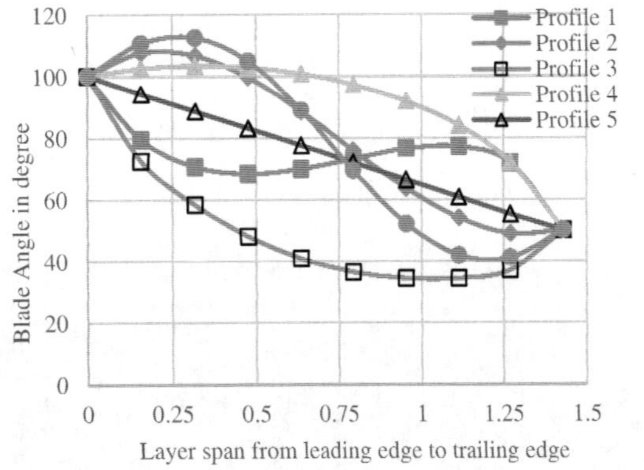

Figure 7: Blade angle distribution from LE to TE

Beside inlet and outlet angle, the shape of the runner is dependent on blade angle distribution from leading and trailing edge. The beta angle distribution for the low head Francis runner is not determined exactly. As shown in figure 7 there are six types of beta angle distribution from the leading edge to trailing edge of the runner. Profile 1 is the rear blade loading profile, which means the decrease in the blade angle initially and gradually increases towards the outlet. Profile 2 is the front blade loading which angle distribution increases at the inlet and gradually decreases at the outlet. Profile 4 is the linear distribution from an inlet to outlet. In figure 8, the blade cross section of 2 and 4 is shown which indicate that the variation of the blade shape with change in the blade angle distribution from an inlet to outlet.

$$\cot\beta_1 = \frac{\pi D_1 B_1}{Q}\left(\frac{\pi D_1^* N}{60} - \frac{60 g H_{th}}{\pi D_1^* N}\right) \tag{9}$$

$$\tan\beta_2 = \frac{Q/A_2}{\frac{\pi D_2^* N}{60}} \tag{10}$$

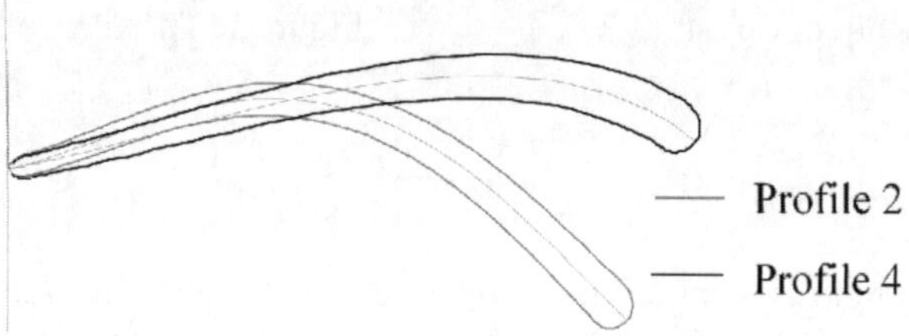

Figure 8: Blade cross section for profile 2 and profile 4 at mid-span

4.2 Thickness Distribution

The thickness distribution of the runner plays the vital role in the strength and performance of the runner. The smooth thickness distribution reduces the friction losses and flow separation in the runner [7]. There is three thickness profile variation, which has been studied and shown in figure 9. The thickness distribution is with positive, negative and neutral variation from leading edge and trailing edge.

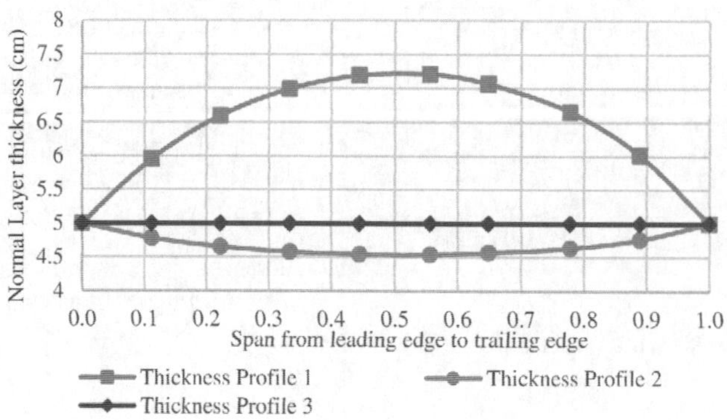

Figure 9: Thickness distribution from leading edge to trailing edge

4.3 Performance

The performance of the runner is affected by the blade angle distribution. The front loading blade profile shows the better performance compare to the other cases as shown in figure 10. The blade angle distribution for strictly increasing and decreasing from the inlet to outlet has the adverse effect on the efficiency due to the swirl flow at the outlet of the runner. The blade angle must follow the front blade loading for the better efficiency and minimization of the swirl flow in the outlet. However, the profile 3 has highly poor performance in comparison to the other blade loading profiles as shown in figure 10. The sudden decrease in the performance of the profile 3 is due to the swirl flow in the draft tube. The flow angle decreases to a minimum value in the mid-span and suddenly start to rise, which is the main cause of the swirl flow at outlet and sudden pressure drop in the mid-span. In profile 6, blade angle after the mid-span is slightly lesser than the outlet angle and it slight effect on the pressure drop at that specific location. Therefore, the blade angle of the runner throughout span should be greater than the outlet blade angle to minimize the swirl flow and have the smooth pressure drop in the blade.

The negative thickness profile has better performance but the low power generation than thicker blade profile as shown in figure 11. Rather than the thicker blade, a thinner blade has the better streamline flow by reduction of secondary flow at the suction side to enhance the performance of the runner by 0.1 %. The pressure drop in the suction side of the thicker blade is less than the thinner one. Overall, the uniform thickness blade performance is better than thicker and thinner profile. The effect of the thickness is not significant to improve the performance of the drastically but slight improvement has been possible from the thickness of the blade.

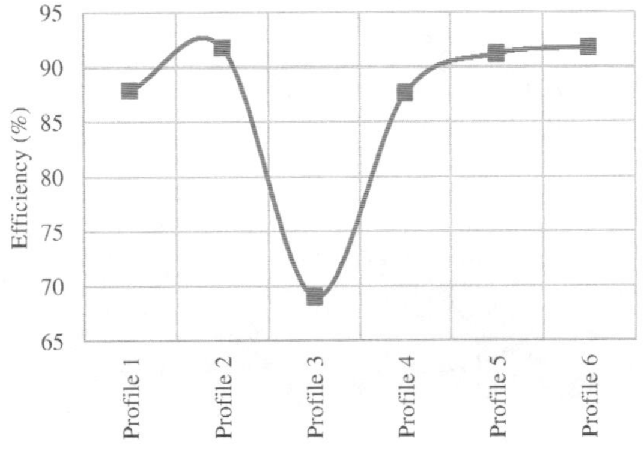

Figure 10: Efficiency versus blade angle distribution profile

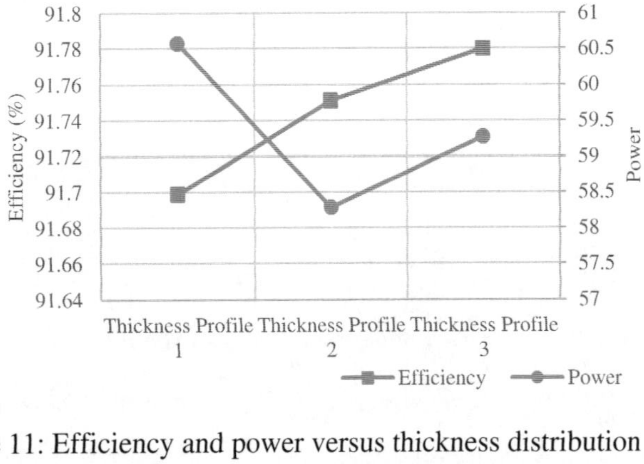

Figure 11: Efficiency and power versus thickness distribution profile

The swirl flow in the draft tube is shown in figure 12, which implies the reason for profile 2 of having better performance. The comparison of the outward flow in the draft tube in figure 12 clarifies that with blade angle distribution such as profile 3 has the most swirl flow because the blade angle at the mid-span is lower than the outlet blade angles. Figure 13 shows the flow angle at the draft tube inlet from the hub to the shroud of the front loading, rear loading, and no loading. There is not any difference in the flow angle at the outlet of the runner in three different loading. However, the circumferential velocity at the outlet of the runner indicates the difference between the three types of blade loading. In an ideal case, the circumferential velocity should be zero at the outlet for the Francis runner for maintaining the uniform flow. Figure 14 indicates that the circumferential velocity at the outlet is approximately zero with the slight deviation in the case of front loading.

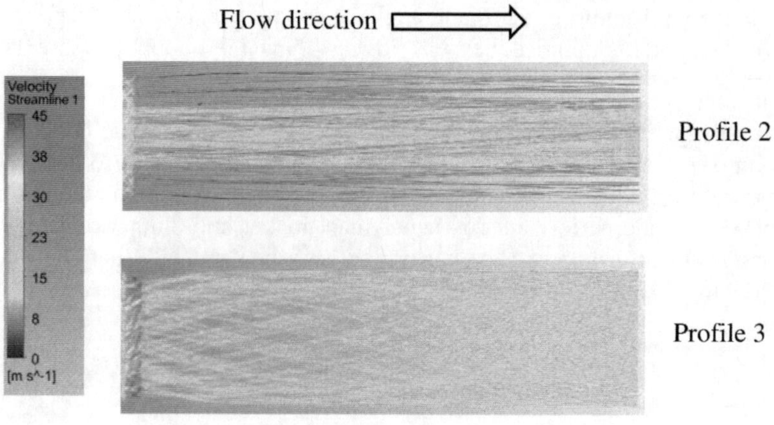

Figure 12: Streamline in the draft tube

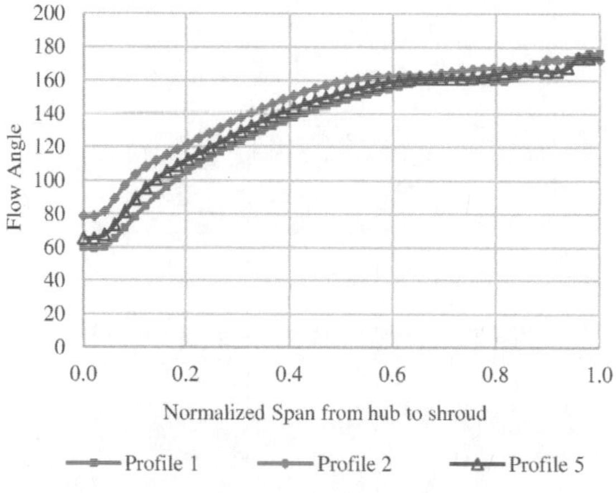

Figure 13: Flow angle versus normalized span

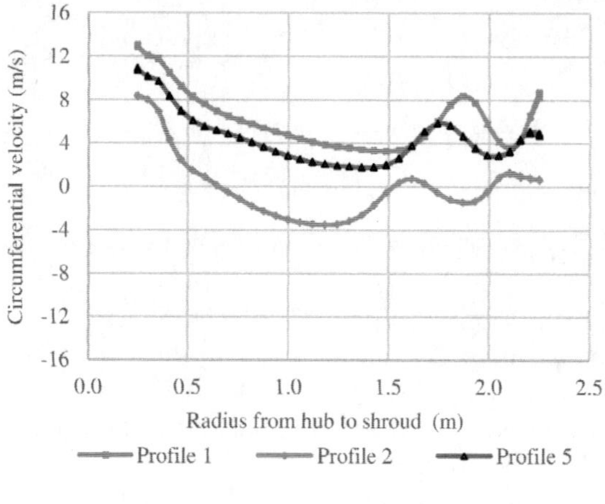

Figure 14: Circumferential velocity versus radius

5. Conclusion

After the studying various blade loading profile, it is concluded that blade loading had the direct impact on the performance. The better performance runner can be designed by following the front blade loading profile. The swirl flow in the draft is suppressed when the blade angle throughout the span is greater than outlet blade angle. When the blade angle at any span is lesser than outlet blade angle causes a change in the curvature of the runner blade causing drastic pressure loss in the runner. Additionally, the better performance and uniform flow in the runner is achieved by front blade loading with blade angle throughout span is greater than outlet blade angle.

Acknowledgement

This work has supported by the New and Renewable Energy of the Korea Institute of Energy Technology Evaluation and Planning (KETEP) grant funded by the Korean Government Ministry of Trade, Industry and Energy (No. 20163010060350).

Nomenclature

ν	Non-dimensional Specific Speed
ω	Angular Speed of Runner
Q	Volumetric Flow Rate
E	Specific Hydraulic Energy
N	Rotational Speed of Runner
β_1	Inlet Blade Angle
β_2	Outlet Blade Angle
D_1	Inlet Diameter of Runner
B_1	Inlet Width of Runner
D_2	Outlet Diameter of Runner
D_1^*	Variable Inlet Diameter with Meridional Plane
D_2^*	Variable Outlet Diameter with Meridional Plane

Reference

[1] Ardanuy J F and Fraile-Mora J J 2006 Variable speed hydro generation: Operational aspects and control *IEEE Transactions on energy conversion* **21** pp 569-574.

[2] Brekke H 2001 Hydraulic turbines design,erction and operation NTNU

[3] Henry P and Hussain Khan M 1997 Francis Turbine Runner: Blade Design in *Hydraulic Design of Hydraulic Machinery*, Wiltshire, Avebury pp 329-372.

[4] Eyup K Salih K Nuri Y and Furkan A 2016 *A numerical case study: Bovet approach to design a Francis turbine runner* 8th International Conference on Sustainability in Energy and Buildings, Turin pp 885-894

[5] ANSYS Inc., "ANSYS Customer Portal," [Online]. Available: https://support.ansys.com/AnsysCustomerPortal. [Accessed 10 1 2018].

[6] Chen Z, Singh P M and Choi Y D 2016 Francis turbine blade design on the basis of port area and loss analysis *Energies* **9** pp 164-171.

[7] Padurean I 2004 Analytical method of the representation of the runner's blade thickness from the Francis turbine in *International Conference on Hydraulic Machinery and Hydrodynamics*

Romania.

IOP Conf. Series: Journal of Physics: Conf. Series **1042** (2018) 012007 doi:10.1088/1742-6596/1042/1/012007

Stress and deformation analysis of double curvature arc dams using finite element method (FEM): A case of budhi gandaki hydropower project

Aanand Kumar Mishra[1*], Ajay Singh[1,2], and Akal Bahadur Singh[3]

[1]*Department of Civil Engineering, Khwopa College of Engineering, Tribhuvan University, Nepal*
[2]*Turbine Testing Lab, Kathmandu University, Dhulikhel, Nepal*
[3]*Department of Civil Engineering, Pulchowk Campus, Tribhuvan University, Nepal*

* *Corresponding author (mishra.ak@khwopa.edu.np)*

Abstract

High rise arc dams are widely used in the development of storage type hydropower project because of the economic advantage. Among different phases considered during the lifetime of dam, control of dam's safety and performance becomes more concerned during the lifetime. This paper proposed the 3 – D finite element method (FEM) for stress and deformation analysis of double curvature arc dam considering the non – linearity of foundation rock following the Hoek – Brown Criterion. The proposed methodology is implemented through MATLAB scripting language and studied the double curvature arc dam proposed for Budhi Gandaki hydropower project. The stress developed in the foundation rock, compressive and tensile stress acting on the dam are investigated and analysed for the reservoir level variation. Deformation at the top of the dam and in the foundation rock is also investigated. In addition to that, stress and deformation variation in the foundation rock is analysed for various rock properties.

Keywords: Double curvature arc dams, Finite element method, Hoek – Brown Criterion.

1. Introduction

In storage type hydroelectric project, dams play important for capacity of hydroelectric project and its economic feasibility. Arch dams are economic and has advantage of higher height to base thickness ratio which differs from other though certain challenges of complexity in analysis and construction. There have been different methods developed for double curvature arch dam analysis [1]. Among different methods, finite element method (FEM) has advantages in handling material in-homogeneity and non-linearity [2]. Safety evaluation of rock foundation problems can be basically categorized in two types as represented by E.Hoek and P.Londe [3], which are

- Rupture of rock or resistance to failure
- Deformations and their effect on foundation rock and the structure

Rupture of rock considers failure of rock mass through over stress of either tensile compressive or shear stress which might be the effect of loads imposed after construction of structure. This can be evaluated using appropriate failure criterion of rock mass. Among different method of analysis of double curvature arch dam three dimensional Finite Element Method (FEM) is adopted for stress-strain analysis in double curvature arch dam along with its abutments and foundation.

2. Project Description and Considerations

The Budhi Gandaki Hydroelectric Project is a storage type of project located between boundary of Gorkha and Dhading district on the Budhi Gandaki River, which is estimated of 1200 MW. Double curvature arch dam having total volume of dam will be 5753000 m^3 of concrete, with total height of dam between deepest foundation level (EL 280m) and the crest parapet (EL 543) is 263m. The dam width at the base is 80m and the developed dam crest length is 760m [4]. Dam is located over bedrock consisting of Nourpul formation. In finite element model, three types of rock are considered, all with similar E-W oriented strike and dip amount 75^0 to north. Apparent thickness of rock layers is considered as, Dandagaun phyllite 580m, Lower Nourpul Formation or Purebesi quartzite 225m, and Upper Nourpul Formation 595m [5].

During finite element analysis, required parameters of bed rock, three formation considered in Hoek–Brown failure criterion is considered as in table 1 and estimated according to Geological, Geotechnical and Seismological Investigations and Studies report of Budhi Gandaki HEP. [4,5]

Other parameters as Young's modulus of elasticity (rock mass) $E_{rm,}$ Hoek and Brown reduced material constant (m_b), Hoek and Brown criterion constant (s and a) and Poisson's ratio of rock mass (ν_m) are determined using following relations [6],

$$\left.\begin{aligned} E_{rm} &= E_i\left(0.02+\frac{1-D/2}{1+e^{(60+15D-GSI)/11}}\right)\\ m_b &= m_i e^{(GSI-100)/(28-14D)};\ s = e^{(GSI-100)/(9-3D)}\\ a &= 0.5+\left(e^{-GSI/15}-e^{-20/3}\right)/6\\ V_m &= 0.5-\frac{0.2RMR}{RMR+0.2(100=RMR)} \end{aligned}\right\} \quad (1)$$

As dam body concrete of M20 grade is considered having modulus of elasticity 20000 MPa, Poisson's ratio 0.2 and unit weight 24 kN/m^3 has been considered [4]. For study of variation in deformation and stress with change in bed rock properties, additional different seven properties of bed rock have been considered with variation in GSI values is considered as shown in table 1 [4,7].

3. Modelling and Analysis

3.1 Finite Element Method

A problem with finite boundaries well, the FEM uses common technique of discretization beyond the zone of influence from the structure. It can be concluded that a foundation model of depth and width equal to 1.5 and 3.0 times the base width of the dam are sufficient to achieve accurate stress results [8].

Applying virtual work method [9], small increment of stress and strain in adopted element is governed by equation (2).

$$\int_{vol}\{\delta\varepsilon\}'\{\delta\sigma\}dv=\int_{areal}\{\delta u\}'\{b\}dv+\int_{\Gamma}\{\delta\varepsilon\}'\{t\}d\Gamma+\sum\{\delta u\}'\{p\} \quad (2)$$

Where, {ε} is the strain vector, {σ} is the stress vector, {U} is the displacements vector, {b} is the body forces vector, {t} is the traction forces vector, {P} is the vector of concentrated forces applied, dv is an element of volume, dΓ is an element of the boundary of the element on which the traction forces{t}are applied.

Using interpolation matrix which also is defined as shape function [N] are represented in equation (3) can be represented as equation (5).

IOP Conf. Series: Journal of Physics: Conf. Series **1042** (2018) 012007 doi:10.1088/1742-6596/1042/1/012007

$$[k]\{a\}=\{f\} \tag{3}$$

Where [k] is elemental stiffness matrix and {f} is elemental force matrix and given by,

$$[k]=\int_{vol}[B][D][B]dv \tag{4}$$

$$\{f\}=\int_{vol}[N]^T\{b\}dv+\int_{area}[N]^T\{t\}dA+\sum[N]^T\{P\}_I \tag{5}$$

For numerical integration of equations (3), (4) and (5) Gauss quadrature integration is used.

3.2 Failure Criterion

For failure criterion of dam body simple Hoek's law is adopted, whereas for bed rock elastic-plastic behaviour is adopted using Hoek and Brown criterion is adopted. In which plastic stress and strain are determined using return mapping method among two popular methods, Euler procedure and return mapping. As presented by Clausen, Johan Christian using Hoek and Brown criterion yield surface equation can be written in equation 5 where $\sigma_1 > \sigma_2 > \sigma_3$ are principle stress tensile stress taken positive and rest of terms are as defined in equation (1) [10].

$$f=\sigma_1-\sigma_3-\sigma_{ci}\left(S-m_b\frac{\sigma_1}{\sigma_{ci}}\right)^a \tag{6}$$

Plastic potential function for non-associated material behaviour for perfectly plastic is chosen similar as yield function, so

$$g=\sigma_1-\sigma_3-\sigma_{ci}\left(S-m_b\frac{\sigma_1}{\sigma_{ci}}\right)^a \tag{7}$$

During return mapping stress after increment of load is obtained by constitutive relationship of deformable solid, and principle stress of stress obtained be σ^B , if it lies out of yield surface i.e. value of gradient $\partial f/\partial\sigma$ than 1 for σ^B , then updated stress σ^c should be computed and slope of line connecting σ^B and σ^C should be equal to slope of plastic corrector which is slope of vector $s=D\times\partial g/\partial\sigma$ then slope of plastic corrector $\alpha_s=s_3/s_1$ and slope of line connecting σ^B and σ^C be α_r, return mapping is expressed in figure 1.[10]

$$\alpha_r=\frac{\sigma_3{}^C-\sigma_3{}^B}{\sigma_1{}^C-\sigma_1{}^B};h_r=\alpha_r-\alpha_s \tag{8}$$

Above equation should value zero and is solved Newton-Raphson method and σ_1^C is obtained by iteration process and σ_3^C is obtained using equation gradient ($\partial f/\partial\sigma$), and

$$\sigma_2{}^C=t_f s_2+\sigma_2{}^B \tag{9}$$

$$t_f=\frac{\sigma_1{}^C-\sigma_1{}^B}{s_1} \tag{10}$$

After updating stress constitutive matrix should also be modified for next increment of load, updated constitutive matrix D^{epc} is obtained using following relations [10].

IOP Conf. Series: Journal of Physics: Conf. Series **1042** (2018) 012007 doi:10.1088/1742-6596/1042/1/012007

$$D^{epc} = D^{C} \frac{D^{C} \frac{\partial g}{\partial \sigma} \frac{\partial f}{\partial \sigma}^{T} D^{C}}{\frac{\partial f}{\partial \sigma}^{T} D^{C} \frac{\partial g}{\partial \sigma}} \tag{11}$$

Where D^c is modified elastic stiffness matrix given by D^c=DT and modification matrix T is given by Plastic strain $\Delta\varepsilon^p$ can be obtained using constitutive matrix and plastic strain.

$$T = \left(I + \Delta\lambda D \frac{\partial^2 g}{\partial \sigma^2} \right)^{-1} \tag{12}$$

A schematic presentation of the Newton-Rapson in the elastic plastic finite element method is presented in the flow chart in figure 3 [10].

3.3 Mesh generation and loads

Hexahedral element with eight nodes is used with total 6807 number of nodes and 5409 number of hexahedral element were developed as shown in figure 2. Boundary conditions are considered by providing zero degree of freedom to bottom most nodes and two degree of freedom is provided to side faces, U/S and D/S faces of model, rest nodes are considered with three degrees of freedom. Total 1219 nodes are completely restrained out of 6807 number of nodes, rest of nodes were assigned 3 degree of freedom which resulted total 17070 degree of freedom in model.

3.4 Loads

Static loading is only considered to contribute for safety stability analysis of wedge in left abutment and Hoek and Brown criterion analysis of arch dam foundation, which are gravity, reservoir water, temperature changes, and silt, ice, uplift, and earthquake loads. Gravity loads due to weight of the material are computed from the unit weight and geometry of the finite elements.

Hydrostatic pressure: The surface loads are then applied to the structure as concentrated nodal loads. Therefore, hydrostatically varying surface pressure can be specified by using a reference fluid surface and a fluid weight density as input. Hydrostatic load is applied for incremental reservoir level as empty reservoir, water level RL 392m (reservoir depth 112m), water level RL 442m (reservoir depth 162m), water level RL 467m (MOL) (reservoir depth 187m), water level RL 492m (reservoir depth 212m) and water level RL 540m (FSL) (reservoir depth 260m).

Silt. According to IS 6512-1984 silt load can be estimated by increasing unit weight of water in silt zone for horizontal pressure by 360 kg/m^3 and vertical pressure by 925 kg/m^3.

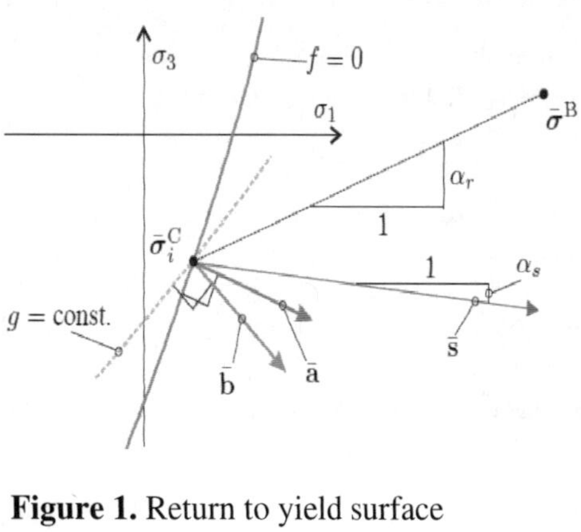

Figure 1. Return to yield surface

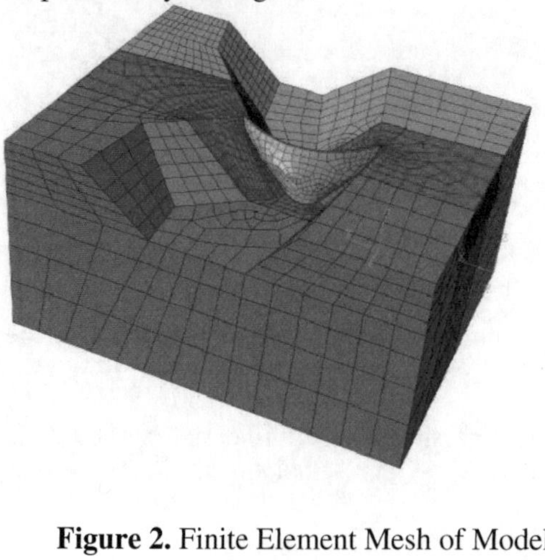

Figure 2. Finite Element Mesh of Model

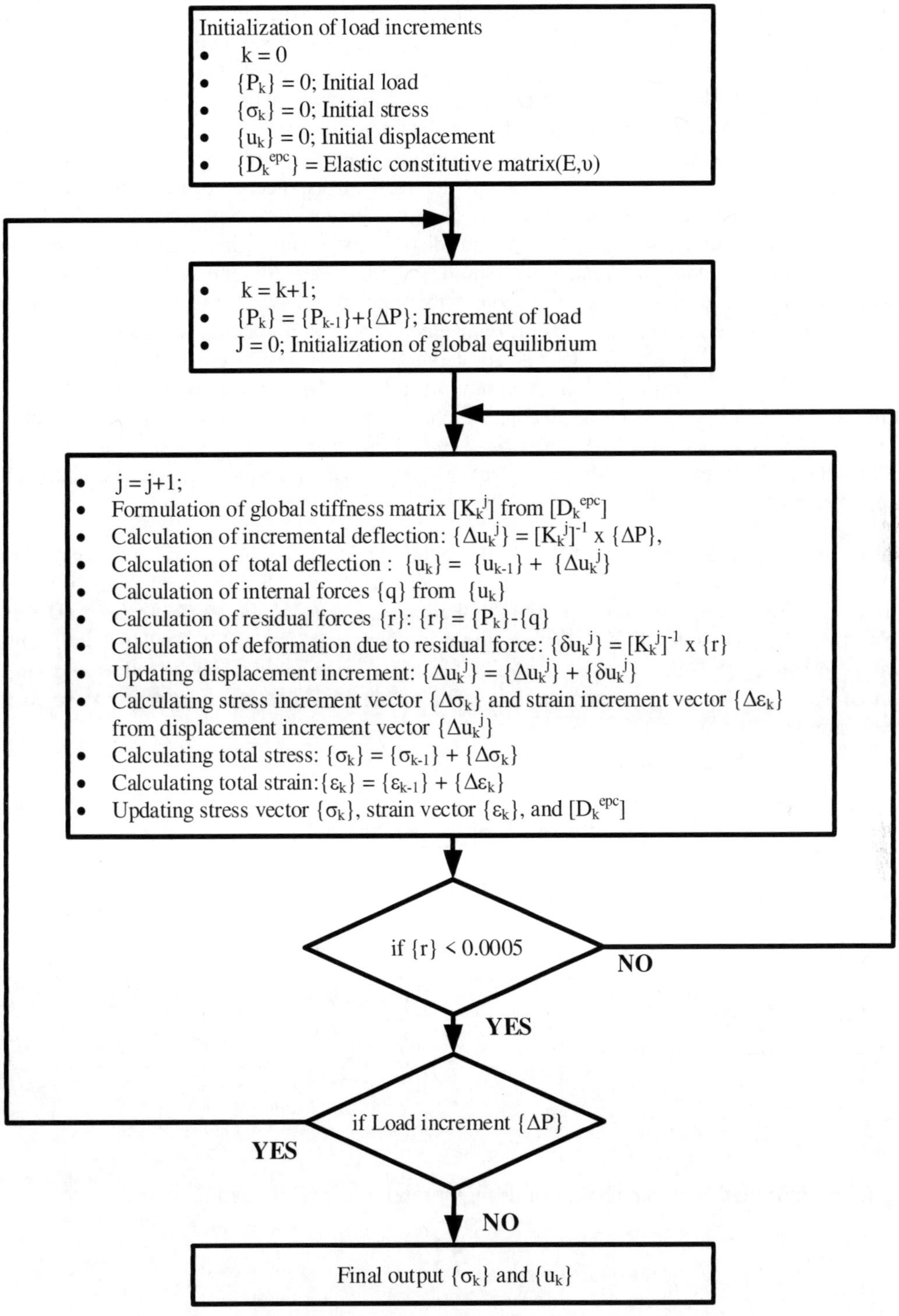

Figure 3. Flow chart for FEM procedure adopted

IOP Conf. Series: Journal of Physics: Conf. Series **1042** (2018) 012007 doi:10.1088/1742-6596/1042/1/012007

4. Results and Discussion

Horizontal deformation at top nodes of the dam at different reservoir level for Case I is presented in figure 4. It is found that D/S deformation of top dam due to gravitational load occurs towards U/S due to curved shape of dam with maximum magnitude of 5.8 cm at mid of dam, with increase in water level deformation gradually shifts towards D/S and maximum deformation is observed 19.4 cm at full supply level. Vertical displacement at top of dam as obtained in figure 5 due to gravitational load only is downward and with increase of water level in reservoir, due to curved shape of dam in vertical axis upward hydrostatic force also acts which reduces the vertical displacement due to gravitational load and with further increase in reservoir level upward displacement has occurred at top of dam, but at bottom of dam there is very less amount of upward displacement although which may produce tensile opening at contact of foundation and dam that should be minimized by increasing base width of dam. Over all deformed shape of dam at empty reservoir level (due to self-weight of dam) and at full supply level is observed as in figure 6 and 7. Total displacement is determined as resultant of vertical and horizontal deformation at different nodes of dam. Contour of total displacement full reservoir level is observed as in figure 8. Principle stress contour for full supply level (maximum static loading condition) are given in figures 9, 10, 11 and 12 which are obtained for case I foundation properties i.e. for actual foundation properties of site. It can be observed that maximum compressive stress developed in dam at full supply level is 11.5 MPa and, tensile stress is developed at contact with abutments (in US face). Which is developed due to bending action in dam due to hydrostatic load.

Results are compared with the report of 'Feasibility Study and Detailed design of Budhi Gandaki HPP' [9]. Variation of the deformation and stress are identical to the 'Engineering guidelines for the evaluation of hydropower projects, chapter 11' by Federal Energy Regulatory Commission Division of Dam Safety and Inspection (1999). Deformation in dam increases with decrease in foundation modulus of elasticity and the increase rate of deformation increases for lesser value of foundation modulus of elasticity, which is also observed. It was also observed that rate of increase in stress is less and found similar to the literature.

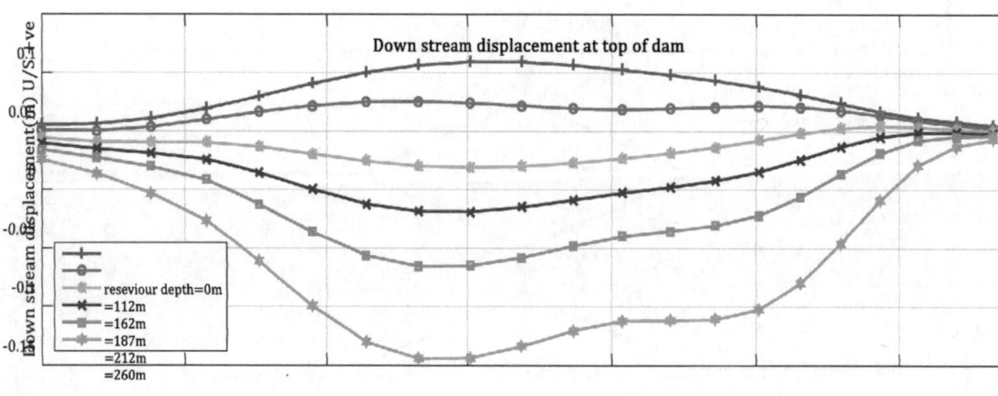

Figure 4. D/S deformation at top of dam for different reservoir level (Case I properties of foundation)

IOP Conf. Series: Journal of Physics: Conf. Series **1042** (2018) 012007 doi:10.1088/1742-6596/1042/1/012007

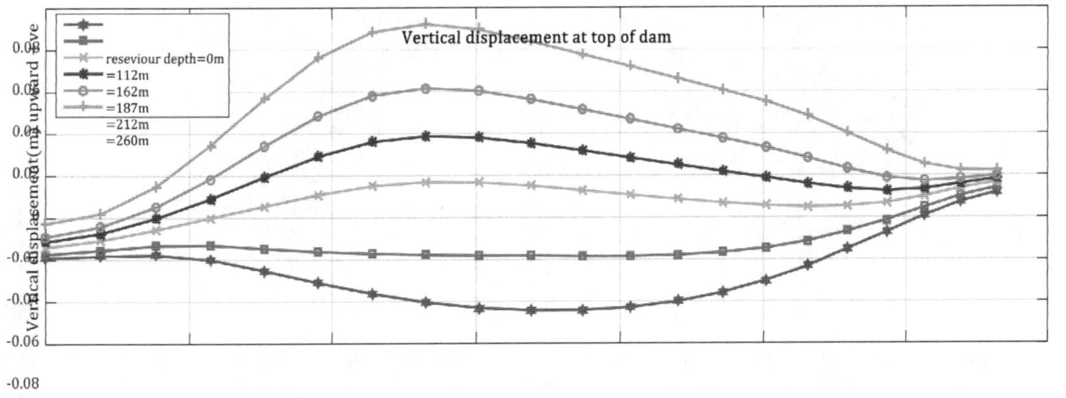

Figure 5. Vertical displacement at top nodes of dam for different reservoir level (Case I properties of foundation)

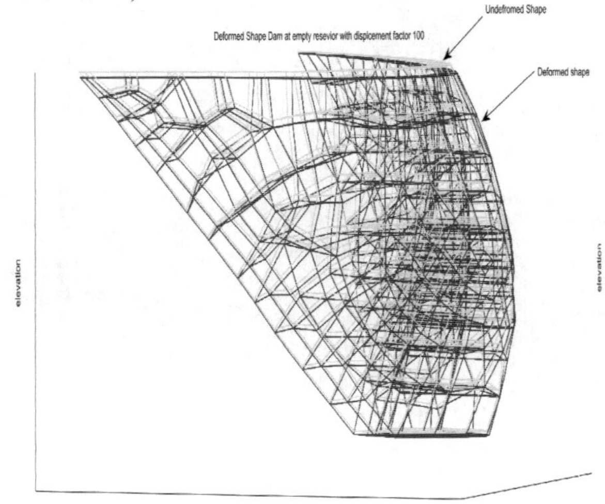

Figure 6. Deformed shape of dam at empty reservoir level (Case I) due to self-weight

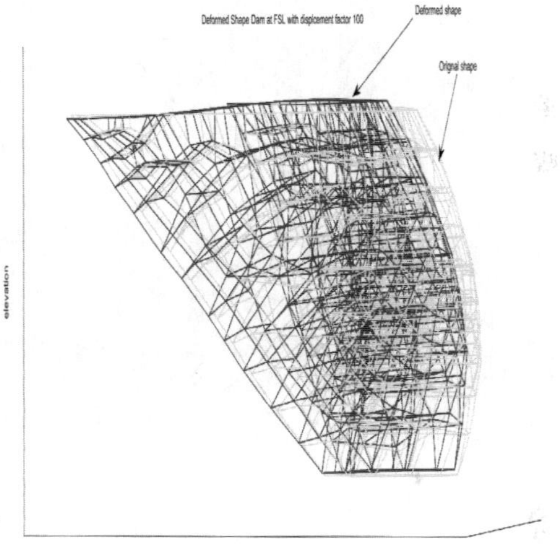

Figure 7. Deformed plot of dam at FSL (Case I)

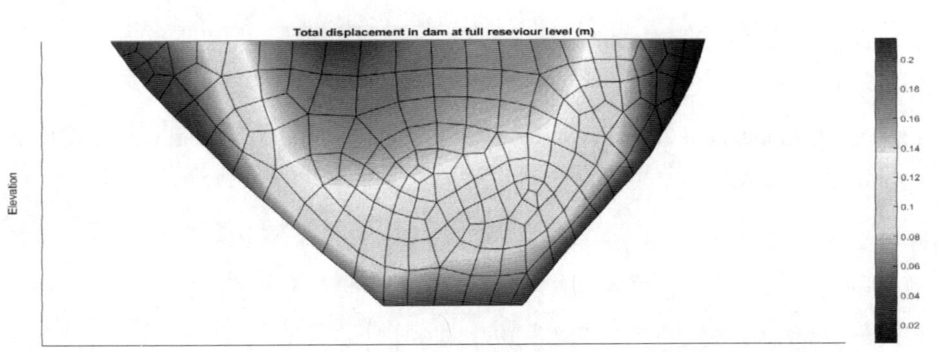

Figure 8. Total displacement (m) contour on U/S face dam at FSL for Case I

IOP Conf. Series: Journal of Physics: Conf. Series **1042** (2018) 012007 doi:10.1088/1742-6596/1042/1/012007

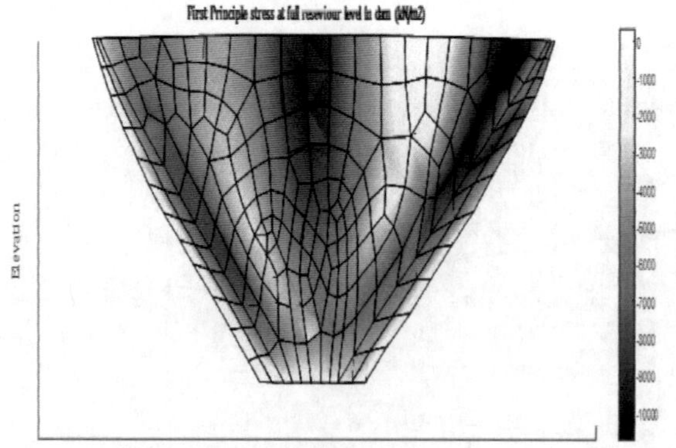

Figure 9. Major principle stress contour on downstream face of dam at FSL

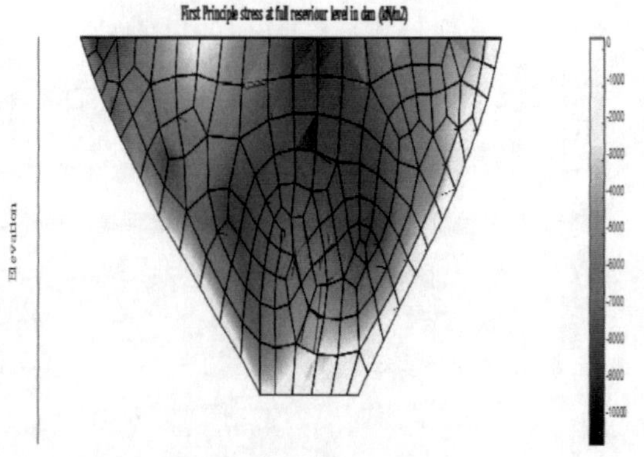

Figure 10. Major principle stress contour on upstream face of dam at FSL

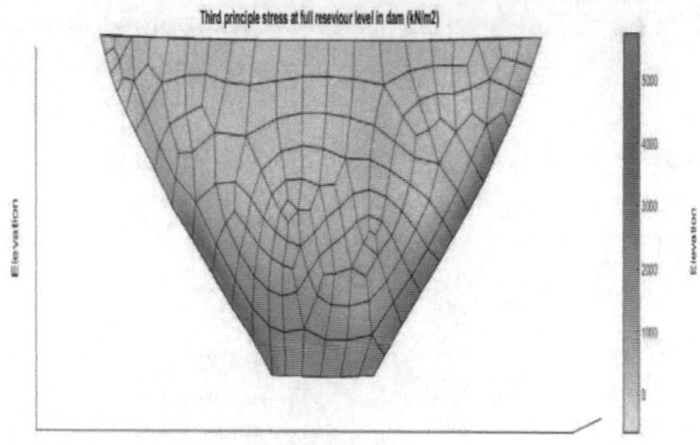

Figure 11. Minor principle stress contour on upstream face of dam at FSL

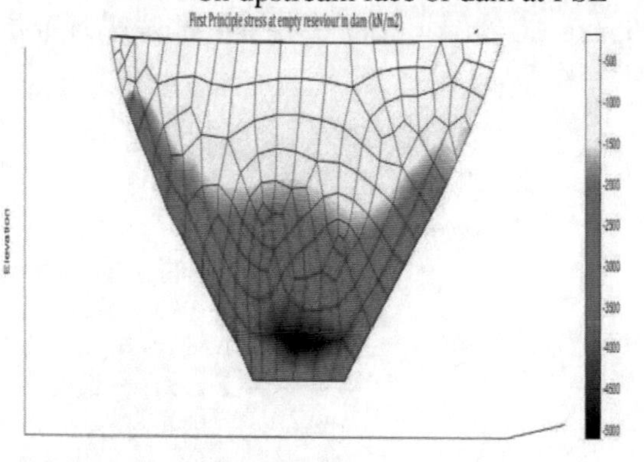

Figure 12. Major principle stress contour on upstream face at empty reservoir

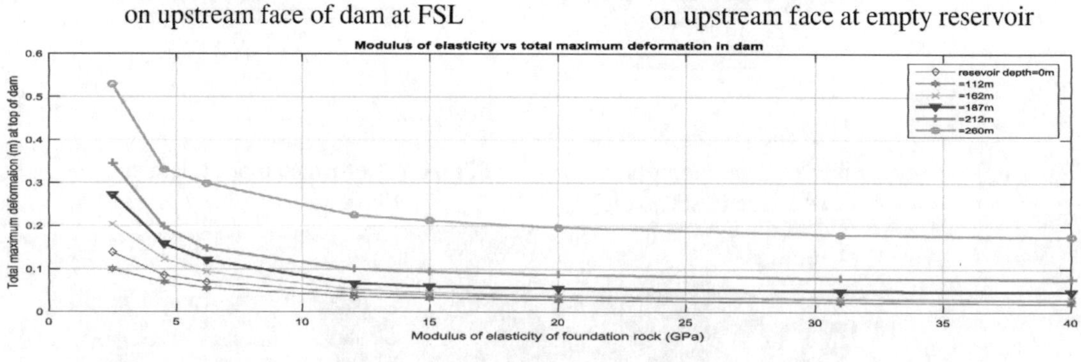

Figure 13. Modulus of elasticity of foundation rock vs Total deformation in dam body

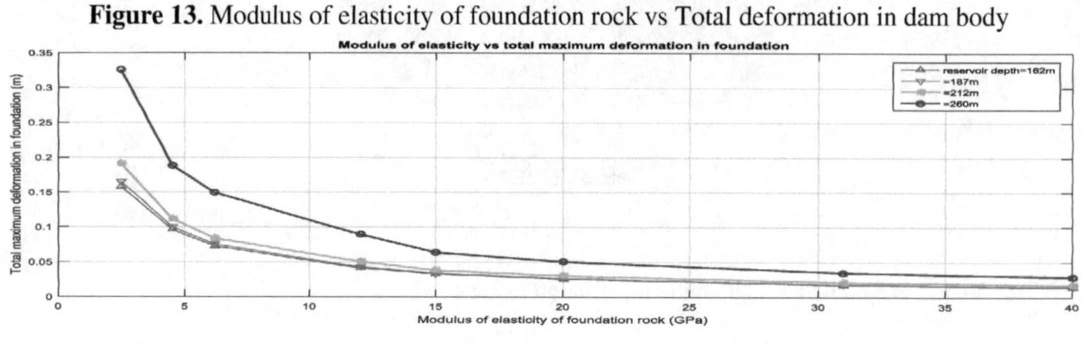

Figure 14. Modulus of elasticity of foundation rock vs Total deformation in foundation

IOP Conf. Series: Journal of Physics: Conf. Series **1042** (2018) 012007 doi:10.1088/1742-6596/1042/1/012007

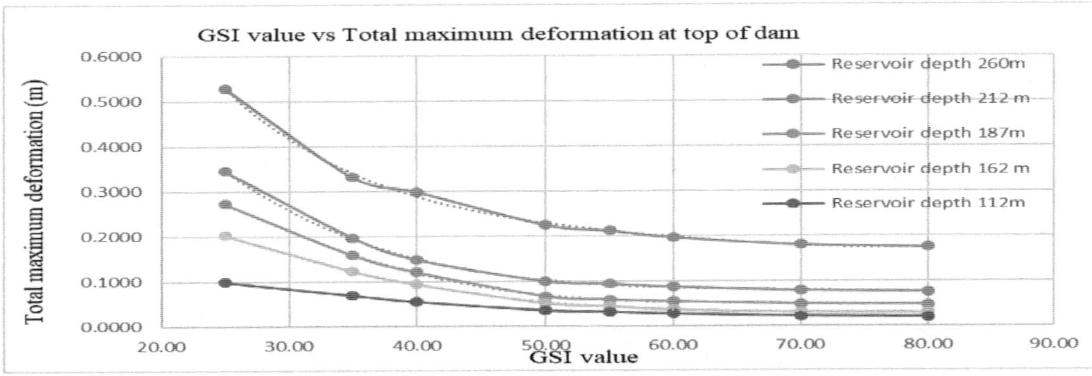

Figure 15. Total maximum deformation in dam vs GSI value of foundation rock

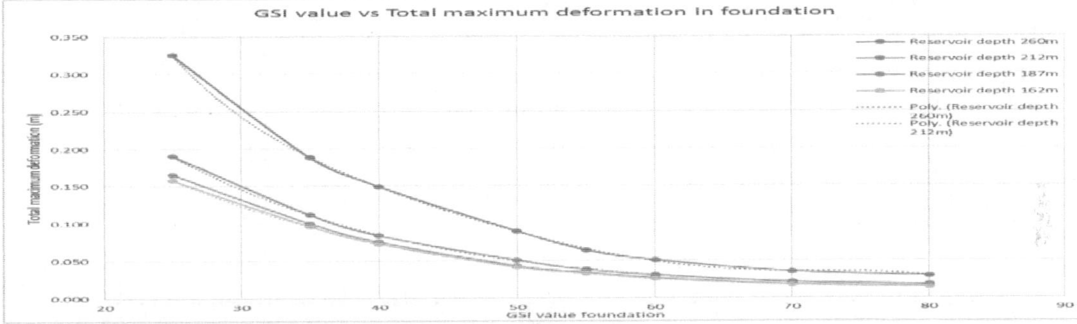

Figure 16. Total maximum deformation in foundation vs GSI value of foundation rock

5. Conclusion

After finite element numerical model analysis results were obtained in terms of nodal displacement, nodal reaction forces, elemental stress, principle stress, strain and plastic strain. Some of these results are presented in previous chapter. At full supply level in reservoir, during this loading, maximum compressive stress in dam and foundation is 12 MPa and 4.2 MPa respectively. Also during full supply level maximum deformation obtained in dam and foundation is 12cm and 6.3cm respectively. From analysis it is observed that with decrease in the GSI value there is significant increase in the deformation in both dam and rock foundation, but there is only slight increment in the stress developed in them.

Rate of variation in deformation in both dam body and foundation not uniform, rate of increment in deformation is lesser for higher value of GSI and rate increases with decrease in the GSI value as represented in figure 15 and figure 16. Similarly, dam body and foundation deformation variation with respect to modulus of elasticity of foundation rock is presented in figure 13 and 14.

Table 1. Diffcrent cases if bed rock properties used for analysis

S.N.	Rock Properties	Case I (Actual Condition of Project)			Case II			Case III			Case IV		
		D.P.	P.Q.	N.P.	D.P.	P.Q.	N.P.	D.P.	P.Q.	N.P.	D.P.	P.Q.	N.P.
1	GSI	40	60	50	30	50	40	30	35	40	40	80	50
2	σ_{ci}(Mpa)	35	62	40	30	62	30	35	62	40	35	62	40
3	E_i(GPa)	40	52	46	40	52	46	40	52	46	40	52	46
4	E_{rm}(GPa)	4.76	15	8	1	12	5.47	1	4.5	5.47	4.76	40	8
5	D	0.2	0.2	0.2	0.2	0.2	0.2	0.2	0.2	0.2	0.2	0.2	0.2
6	m_b	0.92	4.907	1.37	0.62	3.3	0.63	0.62	1.82	0.63	0.92	10.85	1.37
7	s	0.0079	0.00855	0.0026	0.0024	0.0026	0.0079	0.0024	0.00436	0.0079	0.0079	0.092	0.0026
8	a	0.511	0.503	0.506	0.522	0.5057	0.511	0.522	0.516	0.511	0.511	0.5	0.506
9	v	0.32	0.27	0.3	35	0.3	0.32	0.35	0.32	0.32	0.32	0.25	0.3
10	Y_{mb}(kN/m^3	22	25	25	22	25	25	22	25	22	22	25	25
S.N.	Rock Properties	Case V			Case VI			Case VII			Case VIII		
		D.P.	P.Q.	N.P.	D.P.	P.Q.	N.P.	D.P.	P.Q.	N.P.	D.P.	P.Q.	N.P.
1	GSI	40	60	50	40	70	50	40	60	50	40	25	50
2	σ_{ci}(Mpa)	35	62	40	35	62	40	35	62	40	35	62	40
3	E_i(GPa)	40	52	46	40	52	46	40	52	46	40	52	46
4	E_{rm}(GPa)	4.76	6.187	8	4.76	31.5	8	4.76	20	8	4.76	2.47	8
5	D	0.2	0.2	0.2	0.2	0.2	0.2	0.2	0.2	0.2	0.2	0.2	0.2
6	m_b	0.92	2.22	1.37	0.92	7.3	1.37	0.92	10.85	1.37	0.92	0.62	1.37
7	s	0.0079	0.0079	0.0026	0.0079	0.028	0.0026	0.0079	0.00924	0.0026	0.0079	0.002	0.0026
8	a	0.511	0.51136	0.506	0.511	0.5013	0.506	0.511	0.5	0.506	0.511	0.62	0.506
9	v	0.32	0.27	0.3	0.32	0.26	0.3	0.32	0.25	0.3	0.32	0.35	0.3
10	Y_{mb}(kN/m^3	22	25	25	22	25	25	22	25	25	22	25	25

6. References

[1] Bureau of Reclamation 2012 Design of Double-Curvature Arch Dams Planning, Appraisal and Feasibility study, *U.S Department of Interior*

[2] Jing L 2003 A review of techniques advances and outstanding issues in numerical modelling for rock mechanics and rock engineering *Int. Jour. Rock Mechanics Mineral Science* Vol. 40 p 283-353.

[3] Hoek E 1994 Strength of rock and rock masses *ISRM News Journal* Vol. 2 No. 2.

[4] Feasibility Study and Detailed Design of BudhiGandaki HPP *Phase 3: Final Detailed Design Report* Vol 1 Main Report BG-DDR-vol.1-Rev.0.

[5] Feasibility Study and Detailed Design of BudhiGandaki HPP *Phase 2: Field Investigation Report* Vol 1. Main Report BG-INV-P2-C-Rev.0.

[6] Hoek E Brown E T 1997 Practical Estimates of Rock Mass Strength *Int. Jour. Rock Mechanics Mineral* Vol. 34 No. 8 p 1165-1186.

[7] Marinos P Hoek E 2000 GSI: a geologically friendly tool for rock mass strength estimation *Conference proceedings on geotechnical and geological engineering, Melbourne Techonomic publisher* p 1422–1446.

[8] Foster J Jones H W 1994 Procedure for Static Analysis of Gravity Dams Including Foundation Effects Using the Finite Element Method – Phase 1B *Technical Report ITL-94-5, U.S. Army Corps of Engineers, Water ways Experiment Station.*

[9] Amar K 2013 Introduction to finite element analysis using MATLAB and Abaqus

[10] Johan C 2007 Efficient Non-linear Finite Element Implementation of Elasto-plastic for Geotechnical Problems

IOP Conf. Series: Journal of Physics: Conf. Series **1042** (2018) 012008 doi:10.1088/1742-6596/1042/1/012008

Pressure pulsations and hydraulic efficiency at Smeland power plant

V S Ulvan*, J O Kverno and O G Dahlhaug

Department of Energy and Process Engineering Norwegian University of Technology and Science, Trondheim, Norway

* Corresponding author (vegarsul@stud.ntnu.no)

Abstract. Smeland power plant in Norway is experiencing pressure pulsations in their Francis turbine when running above best efficiency point. By measuring both the pressure pulsations and runner efficiency, the cause and effect of the pulsations are to be investigated thoroughly, which is this works main purpose. To find the Francis runners efficiency the thermodynamic method has been used, which builds on the principle that all of the hydraulic losses turns into heat in the flow itself. By measuring the change of temperature before and after the turbine one can, with little other data, calculate the hydraulic efficiency. To identify the pressure pulsations, pressure transducers were placed on the inlet to the spiral casing, draft tube, and upper labyrinth. While doing measurements, air-injection through the runner was tested on full load, which nearly eradicated the pressure pulsations. This might be due to an increase of volume in a pulsating full load vortex that changed its eigenfrequency, and therefore stopped resonating.

1. Introduction

Table 1. Smeland power plant characteristics [4]

Characteristic	Data
Runner	Low head Francis
Head	95 m
Installed power	24 MW
BEP	20 MW
Annual production	119 GWh
Owner	Agder Energi
Built	1985
River system	Mandalsvassdraget

A hydro power plant in Vest-Agder County, Smeland power plant, is experiencing pressure pulsations in their turbine. The pulsations they have registered were of a low frequency variety, 2.8 Hz, which indicated that the the problem may be in the draft tube. The turbine is equipped

IOP Conf. Series: Journal of Physics: Conf. Series **1042** (2018) 012008 doi:10.1088/1742-6596/1042/1/012008

with a check valve which lets air into the draft tube at sub-atmospheric pressures, however this proved to be blocked. The owners of the power plant, Agder Energi, deemed the pressure pulsation matter great enough to seek help from the Waterpower Laboratory at NTNU in order to find out what is causing this phenomenon. By taking measurements of both the hydraulic efficiency and the pressure pulsations one might be able to discern and find the cause of the pressure pulsations.

2. Hydraulic efficiency and the thermodynamic method

In a water turbine, energy is converted from hydraulic energy to mechanical energy which in turn drives an electric generator. However, not all of the energy is converted into mechanical energy, i.e. there are losses through the system. The turbine efficiency says how much of the available hydraulic energy that is turned into mechanic energy. The thermodynamic method builds on the principle that all of the losses in the flow going through the turbine turns into heat in the flow itself. Then the hydraulic losses in a water turbine can be accounted for by the change of temperature in the water running through it, or more accurately the change in enthalpy. By mainly measuring the temperature and pressure at the inlet and outlet, the turbine hydraulic efficiency can be found. The thermodynamic method is a method for measuring efficiency which does not require a direct measurement of the volumetric flow, which can be a very difficult parameter to measure in existing power plants. [1]

The general equation for calculating the hydraulic efficiency is found by dividing the mechanic power by the hydraulic power [3]. The subscripts denotes place of measurement in figure 1.

$$\eta_h = \frac{P_m}{P_h} = \frac{(\rho Q g) E_m}{(\rho Q g) E_h} = \frac{E_m}{E_h} \tag{1}$$

$$\eta_h = \frac{E_m}{E_h} = \frac{\overline{a}(p_{1-1} - p_{2-1}) + g(z_{1-1} - z_{2-1}) + \frac{1}{2}(c_{1-1}^2 - c_{2-1}^2) + \overline{c}_p(T_{1-1} - T_{2-1})}{\frac{1}{\rho}(p_1 - p_2) + g(z_1 - z_2) + \frac{1}{2}(c_1^2 - c_2^2)} \tag{2}$$

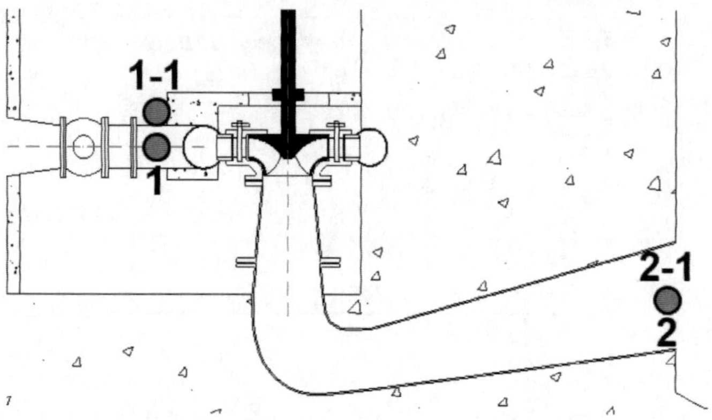

Figure 1. Different places of measurement

IOP Conf. Series: Journal of Physics: Conf. Series **1042** (2018) 012008 doi:10.1088/1742-6596/1042/1/012008

3. Pressure pulsations and full load vortex

One characteristic of hydraulic turbines and pipe flow is pressure pulsations, as the system is dynamic and the flow is often unstable to some extent. These instabilities do however tend to reach a point of equilibrium with the dampening effects of friction as the oscillation amplitude is increased. When a Francis turbine with a fixed rotational frequency operates outside of its design point, the flow leaving the Francis runner will have a rotating component. The direction and magnitude of the rotating velocity component will depend on whether the turbine is operating at part or full load, and how far off the design point it is, respectively. As a swirling flow moves through a cylinder, the bulk of the fluid transport will be along the walls, while a more stagnant region is found at the centre. If the swirl is severe enough, this stagnant flow might stop or move upstream, and a vortex breakdown occurs. [2]

When a Francis runner is operating at full load, i.e. above BEP, a symmetrical vortex appears. Due to the rotation of the flow, the bulk mass flow will occur along the walls of the draft tube, which severely increases the downwards velocity of the water. This vortex can pulsate if the frequency of this pulsation resonates with an exciting frequency from the system. This is when resonance can occur. The exact frequency of this vortex pulsation is difficult to pinpoint but it will change depending on the volume and pressure, as both are parameters that dictates the natural frequency of a gas bubble suspended in a liquid. [2]

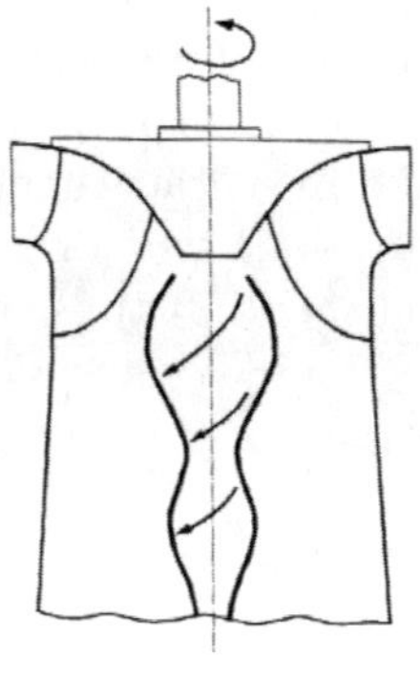

Figure 2. Full load vortex [5]

We can simplify the full load vortex to be a bubble. To find the natural frequency of such a bubble, the problem was simplified and regarded as an analogy to a mass spring system

$$\omega = \sqrt{\frac{k}{m}} \rightarrow f = \frac{1}{2\pi}\sqrt{\frac{k}{m}} \tag{3}$$

where k is the spring constant and m is the mass attached to the spring. k is defined to be

$$k = \frac{dF}{dx} \tag{4}$$

and could, when dealing with changing volumes and pressure, be defined as

$$k = \frac{dp}{dV} \tag{5}$$

$$\Rightarrow dp = \frac{-\gamma p_0}{V_0} dV \tag{6}$$

IOP Conf. Series: Journal of Physics: Conf. Series **1042** (2018) 012008 doi:10.1088/1742-6596/1042/1/012008

where γ is the ratio of specific heats for the gas, p_0 is the pressure, and V_0 is the volume of the gas. Equation 6 stems from the derivative of the equation of state. Professor Nielsen at the Waterpower Laboratory suggested that this, combined with the momentum equation could be used to find the natural frequency of the vortex. He derived the following expression

$$f_e = \frac{1}{2\pi}\sqrt{\frac{\gamma p_0}{V_0 I}} \tag{7}$$

where f_e is the eigenfrequency of the vortex filament and I is an inertial factor related to the mass and inertia of the surrounding water. It has the unit kg/m^4, which Nielsen suggested could be something like ρ/l_c, where ρ is the density of the water, and l_c is some length scale related to the filament. The natural frequencies for various vortex volumes were calculated, with l_c being set to both the circumference, diameter and radius of the vortex cross section, and compared with the frequencies observed at Smeland power plant. Out of all of these, the diameter seems to make most sense in terms of what type of length scale that would affect the flow.

4. Measurement set-up

In order to calculate the efficiency seven sensors were used, five temperature sensors and two pressure transducers. Using a probe on the inlet bleed valve, one temperature and pressure sensor were used to find inlet temperature with corresponding pressure. One temperature sensor measured the leakage water, and the last three measured the temperature in the outlet of the draft tube. The last pressure sensor was placed directly on the inlet pipe in order to calculate the hydraulic specific energy. To find the pressure in the outlet, water column calculations were utilized using the measured height from upper draft tube floor to the water surface.

Pressure pulsation measurement were done with five pressure sensors; one upstream and downstream the main inlet valve, two on the draft tube cone 180° apart, and lastly, one on the upper labyrinth.

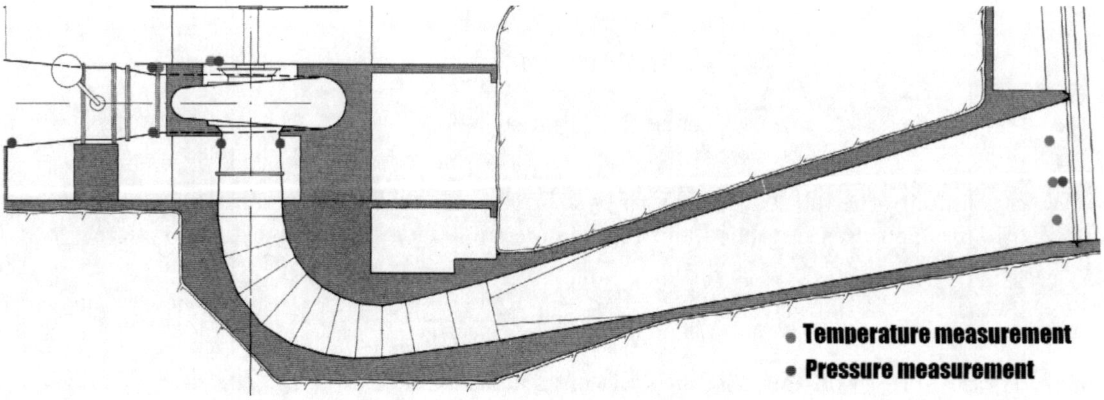

Figure 3. Measurement of temperature and pressure

5. Measurement procedure

Thirteen measurements were made with ten different points of operation. BEP was measured twice due to unstable temperature in the first measurement, 23 MW was done thrice to check repeatability and to test air injection.

Table 2. Sensors

Name	Type	Systematic accuracy
Oceanographic SeaBird 38	Temperature	$0.0001°C$
GE Druck UNIK-5000 3bar a	Pressure	0.01%
GE Druck UNIK-5000 5bar a	Pressure	0.01%
GE Druck UNIK-5000 15bar a	Pressure	0.01%
GE Druck UNIK-5000 50bar a	Pressure	0.04%

Table 3. Points of operation

Measurement #	P_{gen} [MW]
1	19.6
2	21
3	22.2
4	23
5	24
6	6.2
7	9.8
8	13.25
9	15.0
10	16.9
11	19.6
12	23
13 (w/air injec.)	23

6. Results

The pressure pulsations started right after BEP and were present toward maximum effect. The pulsations varied in amplitude and frequency, where both peaked around 23 MW. The results presented here will mainly concern the pulsations at its worst, namely at 23 MW. In the figures, H is the measured head and H0 is net head.

As seen in figure 4, the pulsations have a peak to peak value of 18 % of the head and are dominating the measurements. The pulsations were of a low frequency, about 2.8 Hz.

In discussion with the staff at Smeland power plant it was revealed that the runner had a check valve designed to let air through its center, and it was decided to test air injection through the valve. Upon inspection it was found that the the valve was blocked, not allowing the draft tube to "breath". The valve was removed and a standard compressor for tools was connected and turned on with a pressure of 8-10 bar. This nearly eradicated the pulsations as can be seen in figure 5.

IOP Conf. Series: Journal of Physics: Conf. Series **1042** (2018) 012008 doi:10.1088/1742-6596/1042/1/012008

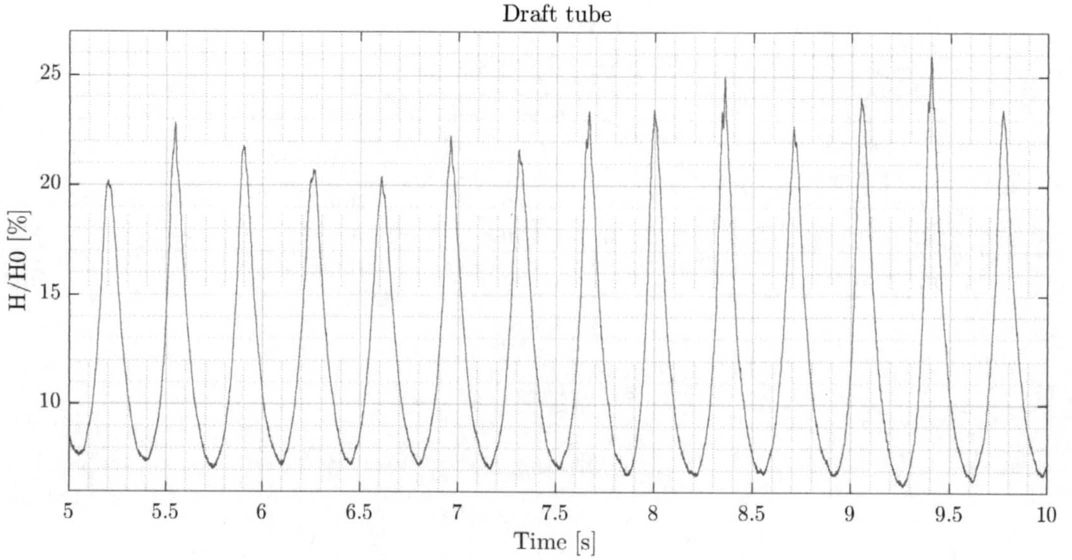

Figure 4. Pressure in draft tube, measurement no. 12

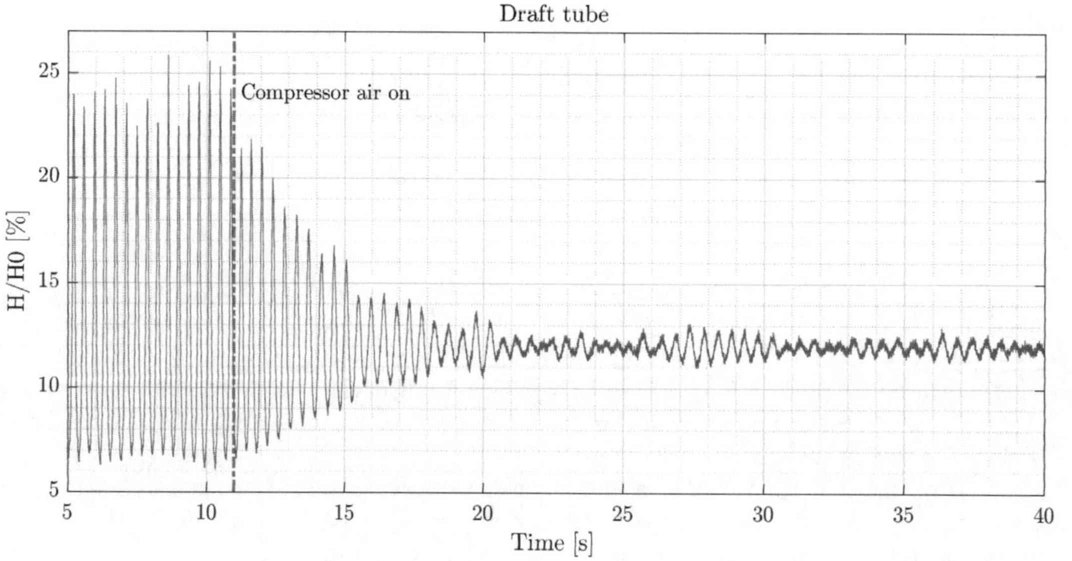

Figure 5. Air injection through the runner, measurement no. 13

As the compressor ran out of air, it's pressure dropped to about 2 bar. Still, the air injection nearly halved the initial pulsations as can be seen in figure 6. It was under these circumstances an additional point for hydraulic efficiency were measured. The efficiency curve can be seen in figure 7, and it is the circle that represent the additional point.

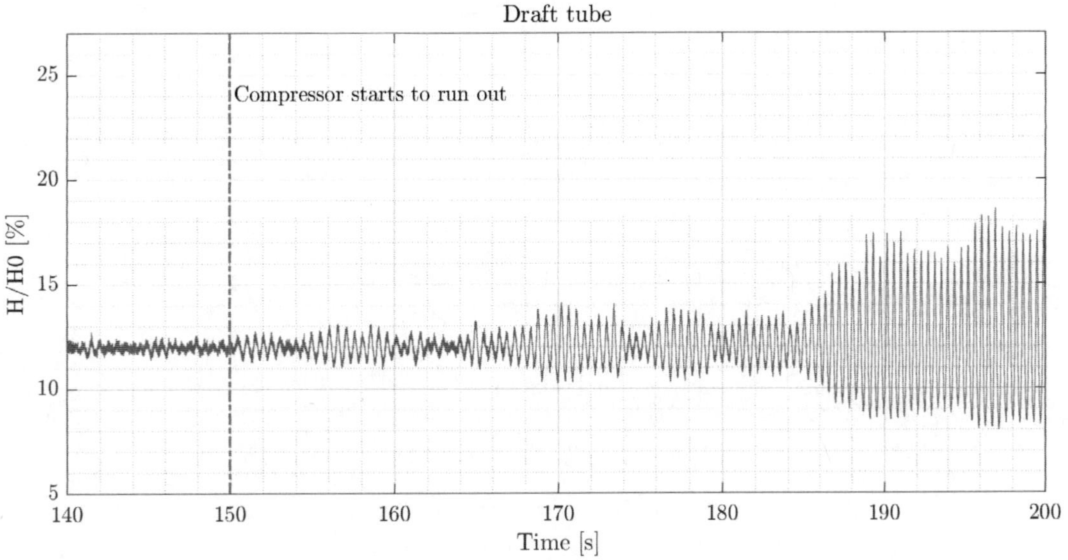

Figure 6. Compressor runs out of air, measurment no. 13

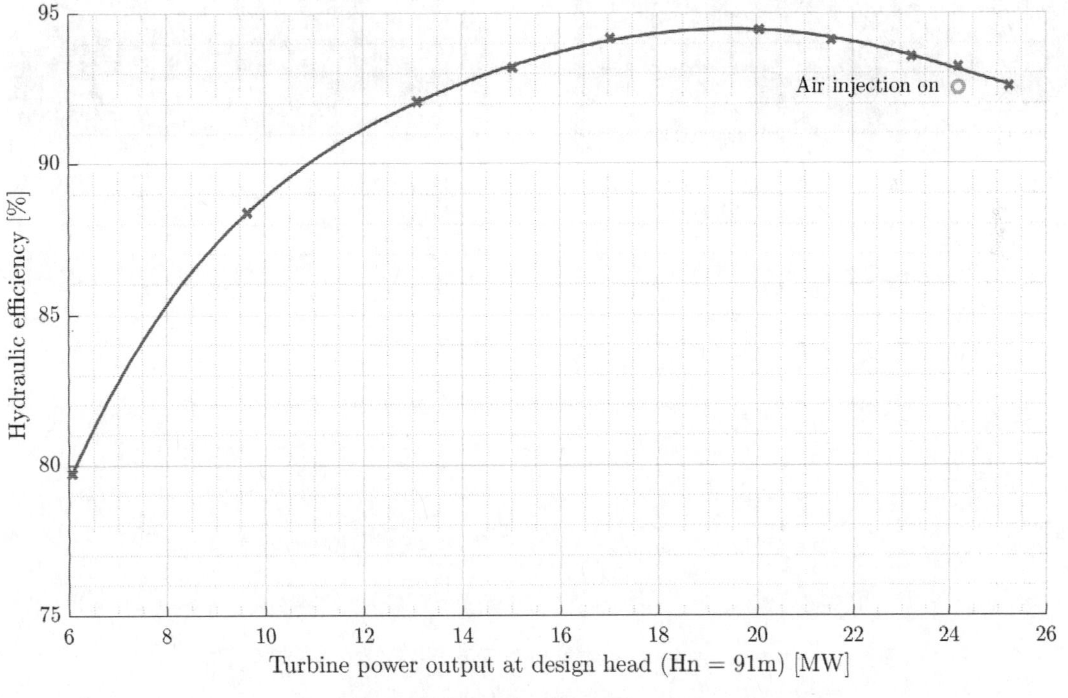

Figure 7. Hydraulic efficiency

An injection of water was also tried, however, as can be seen in figure 8, this had no perceivable effect.

IOP Conf. Series: Journal of Physics: Conf. Series **1042** (2018) 012008 doi:10.1088/1742-6596/1042/1/012008

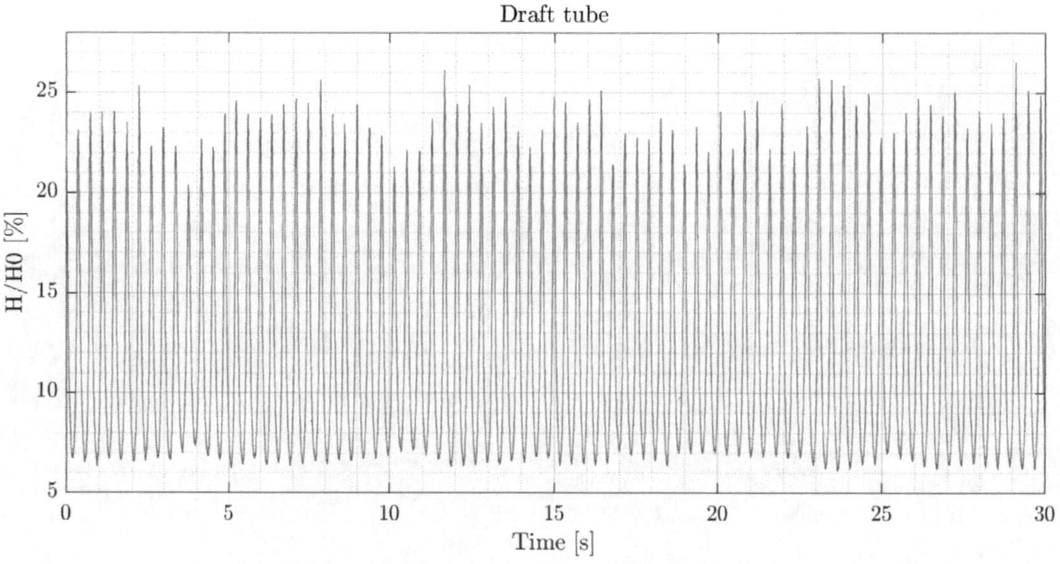

Figure 8. Water injection, measurment no. 13

Figure 9. Staff at Smeland checking valve

Figure 10. Compressor connection

IOP Conf. Series: Journal of Physics: Conf. Series **1042** (2018) 012008 doi:10.1088/1742-6596/1042/1/012008

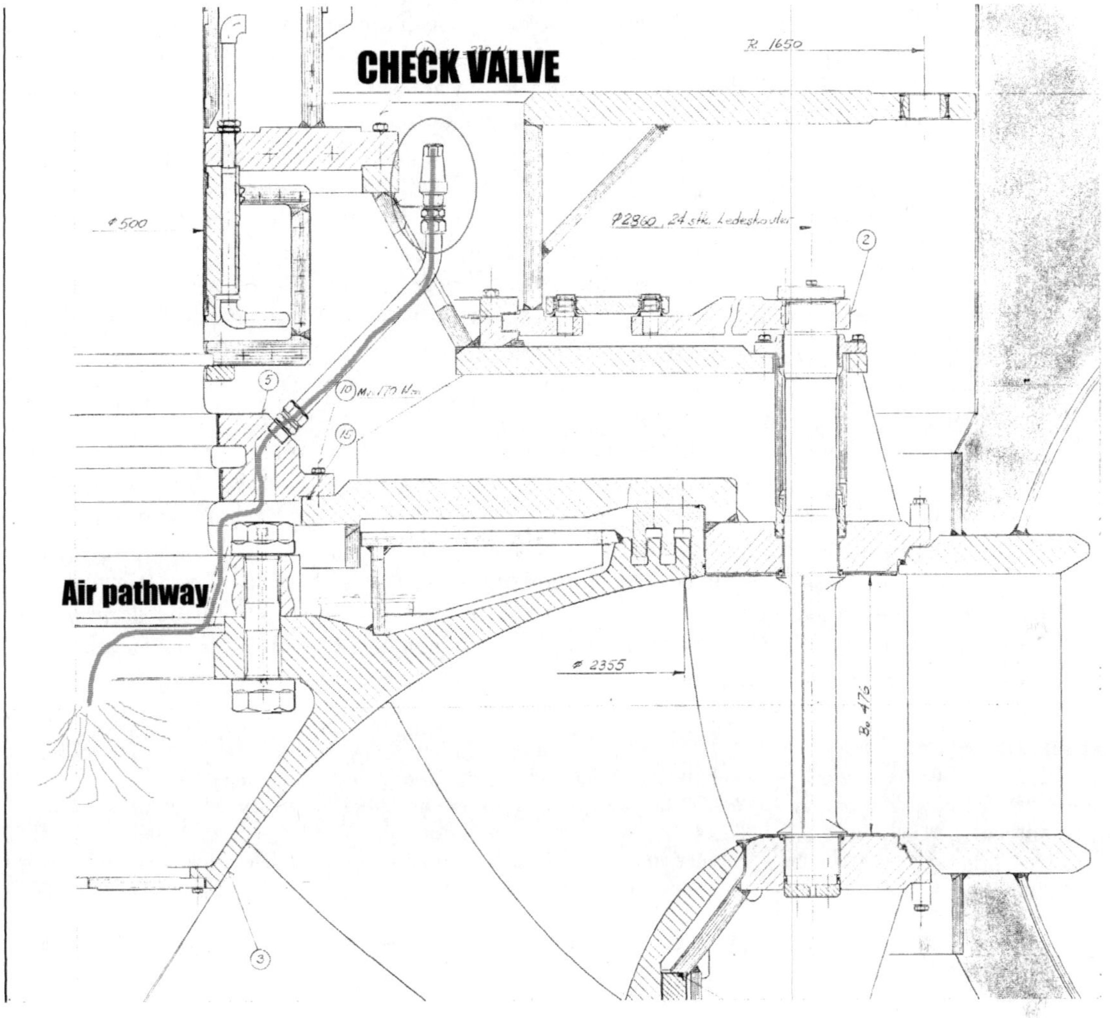

Figure 11. Vertical cross-section of turbine showing check valve and air pathway

7. Discussion

The pressure pulsations at Smeland power plant occur at full load, and the frequency is seemingly dependent on the load. This implies that there is a full load vortex happening in the draft tube, which has is own eigenfrequency. If this eigenfrequency were to match any other pressure pulsations happening in the power plant, resonance would present itself. This other frequency could be a number of many things; pressure wave propagation in the water, von Karman vortices, natural frequency in construction elements, pole passing frequency, RSI, the list goes on.

When injecting air through the runner vertically, with no rotational element, mass in form of gas is essentially added. This would change the volume of the bubble in the draft tube cone, which in turn changes the bubbles eigenfrequency, as the eigenfrequency is a function of volume. Moving the eigenfrequency means leaving the range in which resonance would occur, as all other frequencies would stay the same.

Looking at figure 4, it looks like the pressure pulsations collapse reaching maximum value. If there really is a full load vortex present, and it collapses, then added air could function as a dampener. This dampening effect might be another reason why air injection worked against the pressure pulsations.

IOP Conf. Series: Journal of Physics: Conf. Series **1042** (2018) 012008 doi:10.1088/1742-6596/1042/1/012008

When it comes to the hydraulic efficiency it would seem like the pressure pulsations had no effect at all, as the efficiency is quite close to the measurements done in 1985 by the turbine supplier. One could argue that the pressure pulsations were present from day 1, and therefore the pulsations were included during the very first efficiency measurement.

The drop in efficiency during air injection is difficult to say anything certain about. When the uncertainty analysis is done the three points at 23 MW will surely overlap, and one could not conclude that air injection reduced the hydraulic efficiency. However, the uncertainty is highly determined by IECs strict standards [3], and the fact that the repeatability of the measurement without air injection gave nearly exact results several hours between is cause for concern. On the other hand the temperature of the air can influence the temperature measurement at the outlet. A more thorough look at the data is needed.

8. Further work

There is a lot of measurement data that still need to be addressed and analysed. An uncertainty analysis on the efficiency measurements are essential, and a harder look at the efficiency with air injection is needed. A spectral analysis of the pressure pulsations must be done, and will give a more clear view of all the present frequencies and their peak to peak values. Furthermore, the theory of a full load vortex pulsating with another pulsation needs to be fully investigated.

If possible a real-life simulation of a full load vortex with air injection will be done in the Waterpower Laboraorty NTNU. This might give more insight into the effects of air injection, as the draft tube cone is see-through.

References

[1] Kjoelle, A. 2003. Hydraulisk maaleteknikk. Grunnleggende prinsipper og maalemetoder, 2.
[2] Dorfler, P., Sick, M., Coutu, A., 2013. Flow-Induced Pulsation and Vibration in Hydro-electric Machinery. Springer-Verlag, London
[3] International Electrotechnical Commission. 1991. Field Acceptance Tests to Determine the Hudraulic Performance of Hydraulic Turbines, Storage Pumps and Pump-turbines. International Electrotechnical Commission.
[4] Agder Energi, 'Smeland kraftstasjon', 2015. [Online]. Available: https://www.ae.no/virksomhet/vannkraft/kraftstasjoner/smeland-kraftstasjon/. [Accessed: 12- Sep- 2017].
[5] Brekke, H. 2003. Pumper og turbiner. Vannkraftlaboratoriet NTNU.

Transient Pressure Measurements in the Vaneless Space of a Francis Turbine during Load Acceptances from Minimum Load

R Goyal[1, 2], B K Gandhi[1] , and M J Cervantes[2]

[1] Indian Institute of Technology Roorkee, Roorkee, 247667 Roorkee, India
[2] Department of Eng. Sci. and Math, Luleå University of Technology, 97187 Luleå, Sweden

E-mail: goel.rahul87@gmail.com

Abstract. Increased penetration of solar and the wind impels the designers of the hydroelectric power generation unit to provide more flexibility in operation for the stability of the grid. The power generating unit includes turbine which needs to sustain sudden change in its operating conditions. Thus, the hydraulic turbine experiences more transients per day which result in chronic problems such as fatigue to the runner, instrument malfunctioning, vibrations, wear and tear etc. This paper describes experiments performed on a high model (1.5:1) Francis turbine for load acceptances from the minimum load. The experiments presented in the paper are the part of Francis-99 workshop which aims to determine the performance of numerical models in simulations of model Francis turbine under steady and transient operating conditions. The aim of the paper is to present the transient pressure variation in the vaneless space of a Francis turbine where high-frequency pulsations are normally expected. For this, two pressure sensors, VL1 and VL2, are mounted at the vaneless space, one near the beginning of the spiral casing and the other before the end of the spiral casing. Both are used to capture the unsteady pressure field developed in the space between guide vanes and runner inlet. The time-resolved pressure signals are analyzed and presented during the transient to observe the pressure variation and dominant frequencies of pulsations.

Keywords: Hydropower, Francis turbine, transients, pressure fluctuations, vaneless space, rotor-stator interaction.

1 Introduction

The renewable energy from solar, wind, and hydro have shown greater interest over the last few decades. The electricity generation from solar and wind is not consistent and produces significant disruption in electricity grid network operation [1-2]. This needs to be balanced by a flexible source of electricity generation which is able to maintain the stability of electricity grid network. Hydropower plays an important role in order to maintain the grid stability. The speedy response of hydraulic turbine during peak and base load demand makes it enable to compensate the grid demand during intermittency. Consequently, this has led to the dynamic instabilities in turbine components due to random operation of the turbine to the grid network [2-3]. This has also significantly affected the overall performance of the turbines.

Load variations in a Francis turbine are accompanied by rapid opening and closing of the guide vanes. Flow through the turbine increases with the guide vane openings during load acceptances. The process starts from a steady state condition of the turbine when the guide vanes angle is set to a particular position, such as no load and the minimum load. The minimum load condition of the turbine is generally required to obtain synchronization between turbine and generator and after this; the required steady state condition is achieved. The runner angular speed does not change during load variation process because the turbine is connected to the generator which is rotating at a synchronous speed.

IOP Conf. Series: Journal of Physics: Conf. Series **1042** (2018) 012009 doi:10.1088/1742-6596/1042/1/012009

Experimental investigations are generally performed under steady state and few transients operating conditions. The pressure was measured in the stationary domains, namely vaneless space and draft tube of the turbine during different load variation conditions [4]. The maximum fluctuations in the turbine were associated to the instabilities of rotating and separated flow field in the vaneless space [2-3].

No measurements in the vaneless space during load acceptances from minimum load have been reported. The present measurements focus on the load acceptances operation from minimum load condition of a mode Francis turbine. The measurements are the part of a series of workshop Francis-99 which aim to provide an open access of the complete design and data of model Francis turbine during steady and transient operations (https://www.ntnu.edu/nvks/francis-99). The aim of the present study is to investigate the pressure fluctuations in the vaneless space during load acceptance from the minimum load.

2 Experimental apparatus and procedures

2.1 Model turbine

A high head model Francis turbine was used in the present investigation. The model turbine is a 1:5.1 scaled model of a prototype turbine installed at Tokke Power Plant, Norway. The complete description of the model and prototype turbines is reported in the literature [3]. The model turbine is installed on a test rig which can be operated in both open and closed loop configurations. The test rig is available at Water Power Laboratory (WPL) of Norwegian University of Science and Technology (NTNU), Norway. A schematic of the test rig is presented in Fig. 1. An open loop configuration was used for the present measurements in which the water from the basement can be pumped to the overhead tank. The water from the overhead tank then flowed down to the upstream pressure tank connected to the turbine inlet. The head to the turbine was maintained almost constant by keeping a uniform water level in the overhead tank. The draft tube of the model turbine was connected to a downstream pressure tank, which was open to the air, and the water was released back to the basement. The model turbine is integrated with 14 number of stay vanes, 28 number of guide vanes, a Francis runner with 15 number of splitters (Half-length blades) and 15 number of full-length blades, and an elbow-type draft tube. The discharge to the turbine was measured by a magnetic flow meter which is installed at the inlet pipeline to the turbine. A differential pressure transducer was used to measure the head across the turbine.

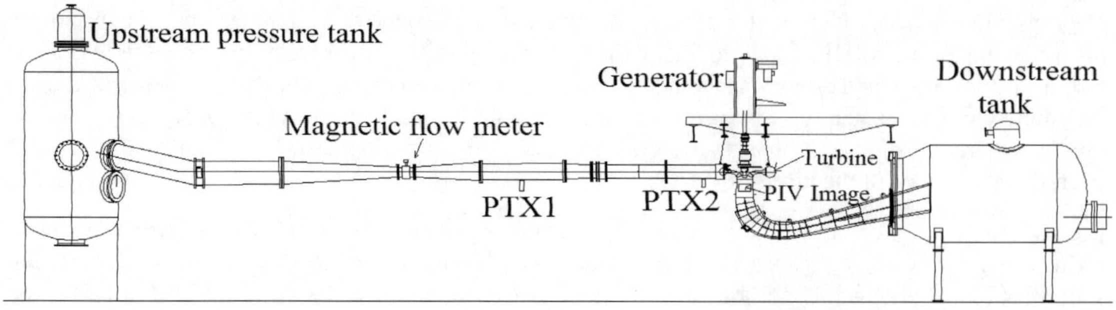

Figure.1 Schematic diagram of model Francis turbine test rig installed at Water Power Laboratory, NTNU.

2.2 Instrumentation and Calibration

The instrumentation and calibration were carried out according to the guidelines available in IEC and ASME-PTC standards on hydraulic turbines [5-7]. The pressure measurements were recorded using a National Instruments (NI) Compact Reconfigurable Input /Output (cRIO) model 9074 ($\leq$ 400 MHz) with a 24 bit, ±60 V analog to digital converter (ADC). The data were sampled at 5 kHz with a separate ADC for each channel. The operating flow parameters such as discharge, inlet and differential pressures, atmospheric pressure, the angular speed of the runner, shaft torque to the generator, bearing friction torque, turbine axial force, and guide vanes angular position were acquired simultaneously through the same data acquisition system. The discharge was measured using a magnetic flow meter (KROHNE IFS 4000 series). In addition to the base instrumentation, two sensors were mounted in the vaneless space, one near the beginning of the spiral casing and one near the end as shown in Table 1 and Figs. 2 (a-b). The radial distance of the sensors in Table 1 is made dimensionless by dividing with the runner radius (r =174.5 mm). The uncertainty of the instruments used in the measurements was determined according to the guidelines available in IEC 60041/60193 [5-7]. The uncertainties of the calibration of the pressure sensors along with the accuracy, as provided by the manufacturer, are described in the literature [8-10]. The total estimated uncertainty was ±0.15% in the hydraulic efficiency at best efficiency point (BEP).

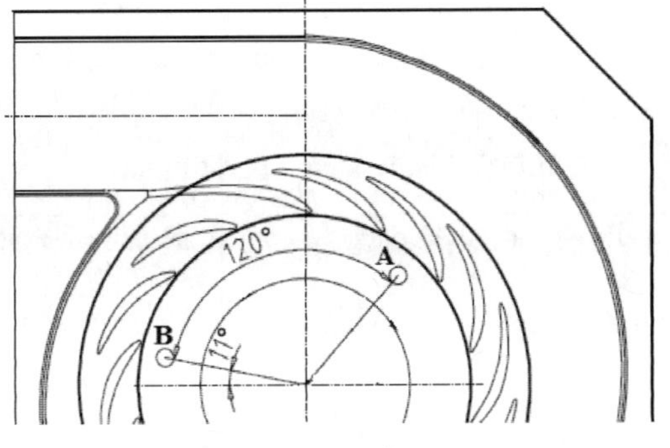

(a) Top view

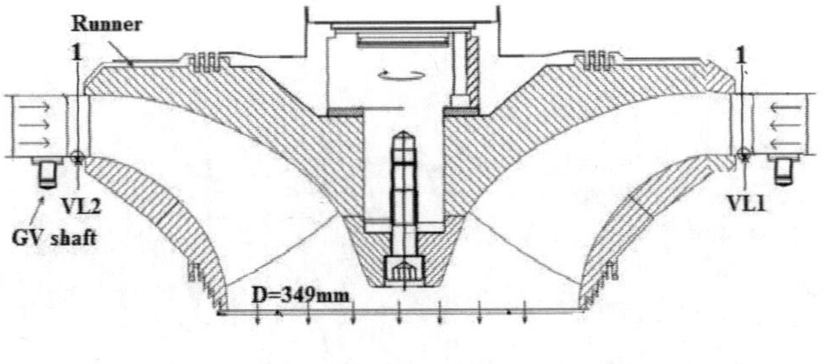

(b) Side view

Figure 2. Sensor placements in the model Francis turbine,(a) Top view, (b) Side view. The positions 1A and 1B are the placements of pressure sensors VL1 and VL2.

Table 1 Position of the pressure sensors

Sensor	Placement	Dimensionless Radial Distance	Type	Uncertainty
VL1	1A	1.23	Kulite	0.01%
VL2	1B	1.84	Kulite	0.01%

2.3 Measurement program

The operating conditions for the measurements were inspired by the operating conditions presented in Francis-99 (I), but with some changes. The transient measurements were preceded and followed by corresponding steady conditions on a model Francis turbine. In all measurements, the runner speed in revolutions per minute (rpm) i.e. 333, was maintained constant to achieve the realistic operating conditions as that of the prototype. The maximum hydraulic efficiency (92.4 %) of the model turbine was estimated at guide vane angle (α) = 9.8°, speed factor (n_{ED}) = 0.179, and discharge factor (Q_{ED}) = 0.15, which was marked as BEP. The modified part load (PL) and high load (HL) conditions were used in order to get the similar runner speed as that of the minimum load. The measurements were performed in an open loop configuration to get the realistic condition without significant variation in effective head available to the turbine. The specifications of the steady state operating conditions used are presented in Table 2.

Table 2 Steady state operating points

Guide vane angle α (°)	Head H (m)	Flow rate Q (m³/s)	Speed factor n_{ED} (-)	Discharge factor Q_{ED} (-)	Hydraulic efficiency η (%)
0.8	12.14	0.02	0.18	0.01	20.9
6.7	11.87	0.14	0.18	0.11	90.1
9.8	11.94	0.20	0.18	0.15	92.4
12.4	11.88	0.24	0.18	0.18	91.8

The speed factor and discharge factor in Table 2 can be computed as:

$$Q_{ED} = \frac{Q}{D^2\sqrt{E}} \quad (-)$$

$$n_{ED} = \frac{nD}{\sqrt{E}} \quad (-)$$

where, Q is the discharge in kg m^{-3} , D is the reference diameter of runner in m, E is the specific hydraulic energy in J kg^{-1} , n is the angular speed of runner in s^{-1}.

Transient measurements were performed by changing the angular positions of the guide vanes from the minimum load (0.8°) to PL (6.7°), BEP (9.8°) and HL (12.4°). The angular position of guide vanes was measured by the voltage signal from the servomotor mechanism. A LabVIEW program with linear coefficients of voltage signals was made to control the exact movement of the guide vanes. A linkage mechanism with servomotor was used for the movement of the guide vanes. The uncertainty of ± 0.044° in recorded positions of the guide vanes was always available because guide vanes were acting against the hydrodynamic forces.

3 Results and Discussion

The obtained results of flow variables (*H, Q, α,* and T_{GEN}) and vaneless space pressure measurements for transients are presented and discussed. Detailed investigations of load acceptance from minimum load are presented with special emphasis pressure fluctuations and frequency content, i.e., rotor-stator interaction and standing pressure waves.

3.1 Flow variables

The representation of turbine load acceptances from the minimum load (ML) with a variation of basic flow variables such as the head (*H*), discharge (*Q*), torque (T_{GEN}), and guide vane position (α) of model Francis turbine are shown in Figure 3. The values α, Q, H, and T_{GEN} in figures were normalized with the maximum value corresponding to 1 (100%) and a minimum value corresponding to 0 (0%) taking the uniform length of the signal using Eq.1.

$$X[-]=\frac{X-X_{Min}}{X_{\max}-X_{Min}} \tag{1}$$

The generator in the WPL was connected to a direct current (DC) converter. The converter unit supplies and takes power to and from the generator so that the selected operating point could be maintained. The converter also fixes the rpm of the generator at the selected value of the turbine runner. In the present study, the generator was rotating at a constant angular speed of 333 rpm which was the synchronous speed. During the measurements at load acceptances, the guide vanes of the model Francis turbine were opened up to the required operating condition. The average angular speeds (ω_{gv}) of the guide vanes movement during the measurement was 0.0022 rad/s.

The transient variation of discharge and generator torque during three different load acceptances process almost follow the trend of guide vanes movement as shown in Figs. 3 (a-c). The maximum output torque from the generator was 9.4, 457.2, 695.9 and 791.4 N-m at ML, PL, BEP and HL operating conditions, respectively. The torque fluctuations (Fig. 3b) for the opening of the guide vanes at BEP were attributed to the oscillations in runner angular speed. The trend of head variation during all transients was almost similar. The head was the most effective parameter during the load acceptances. Initially, a sudden drop in head values was observed up to 3-4 s after the transient. Thereafter, head starts to increase and settles around 11 s after the guide vanes stop. The range of the head variation during the load acceptances from ML was 11.76 to 12.64 m.

IOP Conf. Series: Journal of Physics: Conf. Series **1042** (2018) 012009 doi:10.1088/1742-6596/1042/1/012009

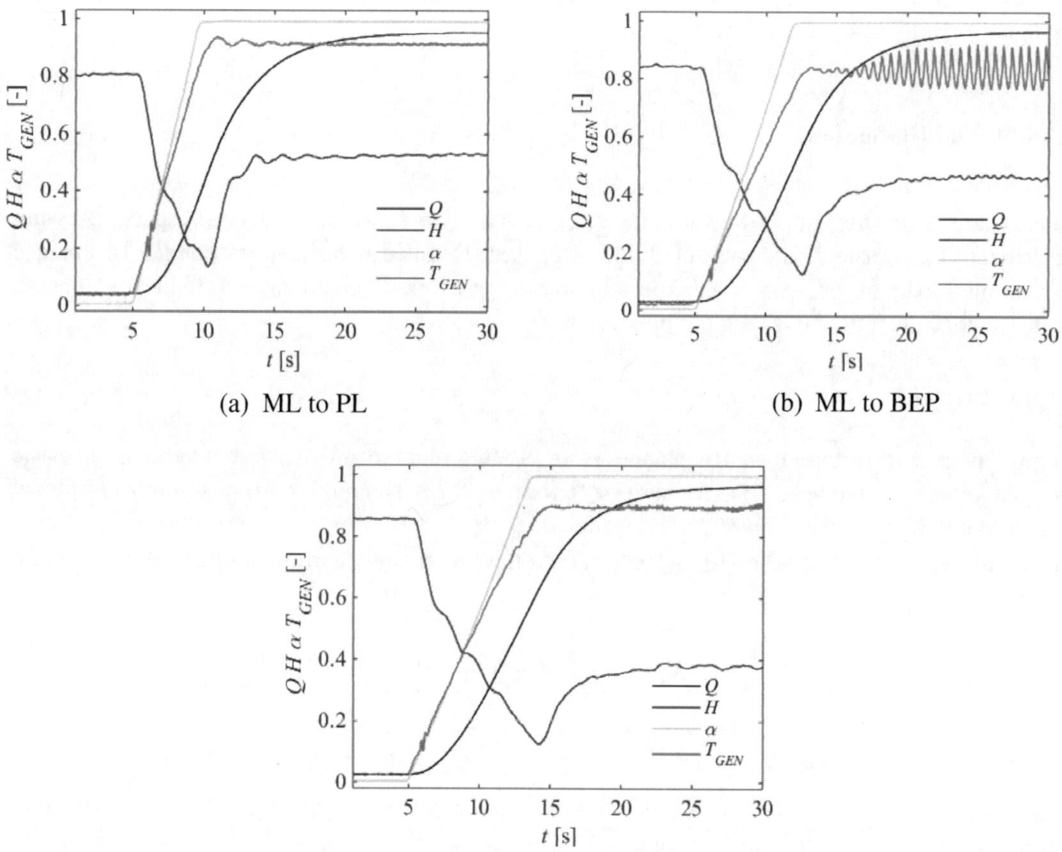

(a) ML to PL

(b) ML to BEP

(c) ML to HL

Figure 3. Transient variation in normalized head (*H*), discharge (*Q*), guide vane angle (α), and generator torque (T_{GEN}) during turbine load acceptances from minimum load; (a) Minimum load to part load, (b) Minimum load to best efficiency point, (c) Minimum load to high load.

3.2 *Pressure fluctuations*

The time-dependent pressure fluctuations during load acceptances from a minimum load of the turbine are shown in Fig. 4. The two sensors, namely VL1 and VL2, located in the vaneless space were selected to show the pressure variation in the turbine. The values of guide vane angle (α) and pressure (P_{norm}) in Fig. 4 was normalized using Equation (1). During load acceptance from minimum load to part load, the overall pressure in the vaneless space was observed to increase from 148 kPa to 163 kPa (9 %) and the transient pressure fluctuations in the vaneless space (VL1 and VL2) was observed to decrease by ± 2.1 % after the transient as shown in Fig. 4 (a). The pressure fluctuation in the vaneless space was attributed to the wakes captured in the vaneless space by the interaction of runner blade and guide vanes.

During load acceptance from minimum load to best efficiency point, the overall pressure in the vaneless space was observed to increase from 148 kPa to 167 kPa due to rapid opening of the guide vanes as shown in Fig. 4(b). The transient pressure fluctuation in the vaneless space was decreased by ±0.7%. The absolute pressure in the vaneless was observed to increase from 148 kPa to 178 kPa during the load acceptance from the minimum load to high load as shown in Fig. 4 (c). The transient pressure fluctuation in the vaneless space was decreased by ± 1.1 %. The different cycles of pressure signal fluctuation in the vaneless and draft tube (Fig. 4) show the existence of high and low-frequency oscillations in the draft tube which is discussed later in the next sections.

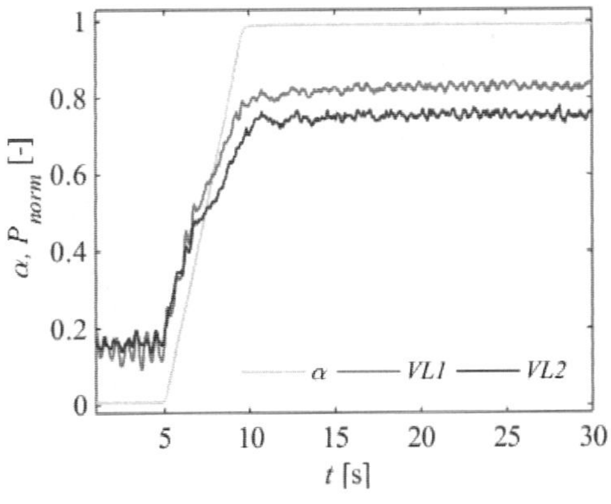

(a) ML to PL

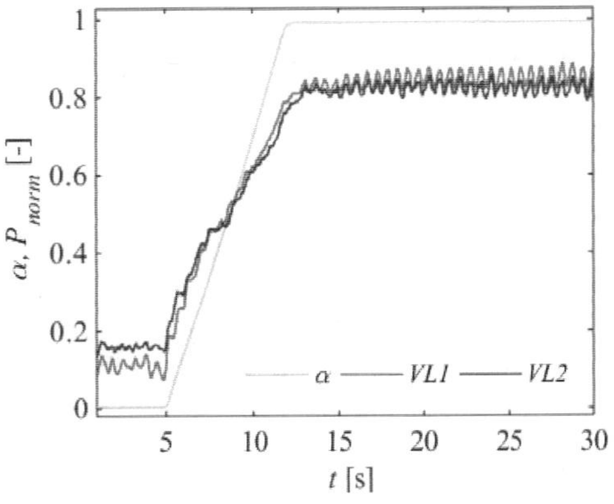

(b) ML to BEP

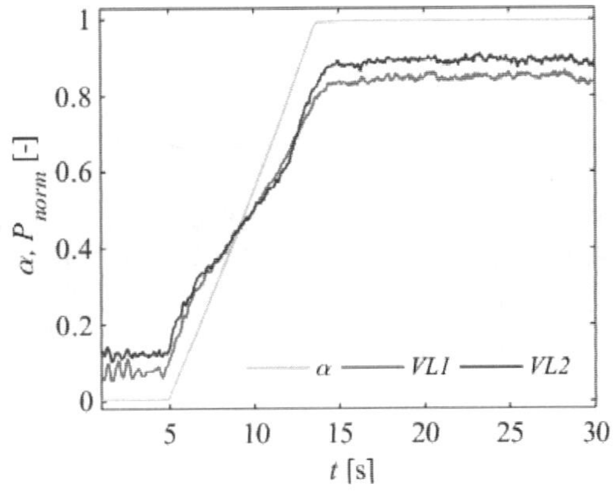

(c) ML to HL

Figure 4. Transient variation in normalized pressure at sensor location VL1 and VL2 during turbine load acceptances from minimum load; (a) Minimum load to part load, (b) Minimum load to best efficiency point, (c) Minimum load to high load.

3.3 Spectral analysis

Spectral analysis of pressure time signals was carried out to investigate the dominant frequencies of the fluctuations in the vaneless space of the turbine. Welch's method with Hanning window on the fluctuating part of the pressure and velocity data was used for the spectral analysis. A built-in function of MATLAB spectrogram was used to present the time-dependent spectral analysis. The function works on the Goertzel algorithm. The window size of 4 s and 2 s with 95% overlap was selected for the spectral analysis of the pressure time signals. The color bar in the spectrograms shows the power spectral density (PSD) of the frequency analysis in logarithmic scale. Equation (2) was used for the PSD analysis, presented in the spectrograms. The spectrograms for the load acceptances from minimum load are presented and discussed.

$$PSD_{\log} = 10 \times \log\,(10 \times PSD) \qquad (2)$$

IOP Conf. Series: Journal of Physics: Conf. Series **1042** (2018) 012009 doi:10.1088/1742-6596/1042/1/012009

The PSD analysis of the pressure signals VL1 and VL2 (see Table 1 and Fig. 2) during load rejection and acceptances is presented in Figures 5 to 7. PSD shows the strength of the corresponding properties (pressure) as a function of frequency. It shows the strong and weak frequency variations in any signal. The y-axis in the Figs. 5 to 7 is normalized using the runner rotational frequency (f_n) of 5.55 Hz.

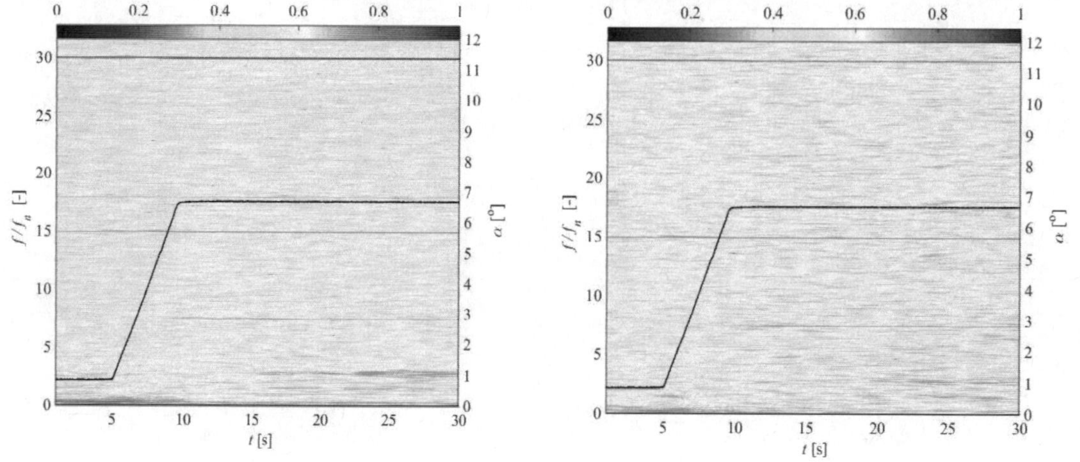

(a) VL1 (b) VL2

Figure 5. Spectrograms of pressure time signals during load acceptance from minimum load to part load; (a) VL1, (b) VL2. Black solid line is the guide vanes opening with y-scale to the right.

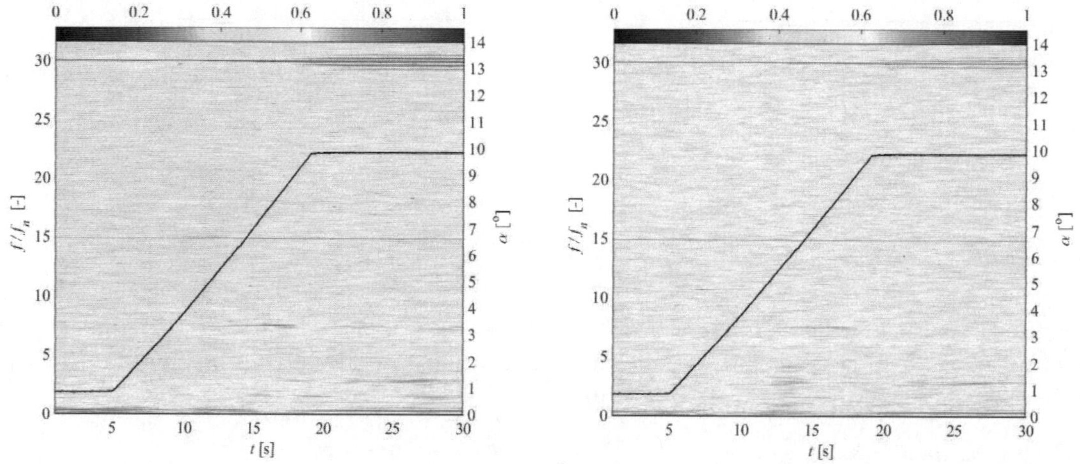

(a) VL1 (b) VL2

Figure 6. Spectrograms of pressure time signals during load acceptance from minimum load to best efficiency point; (a) VL1, (b) VL2. Black solid line is the guide vanes opening with y-scale to the right

The spectrograms of the all load acceptance cases show that for the pressure signals at VL1 and VL2, the normalized blade passing frequency (30) is present in the turbine as shown in Figures 5 to 7. The harmony (15) of the blade pass frequency is also observed in the vaneless space during load acceptances. The PSD strength of the blade pass frequency was observed to increase with the higher load operating condition of the turbine. The maximum strength was observed at HL operating condition of the turbine due to reduced vaneless space between guide vanes trailing edges and runner leading edges. Seen in spectrograms, the pressure signal at location VL1 shows more PSD strength as compared to that of the VL2. This may be attributed to higher flow rate near VL1 location which is located near the beginning of the spiral casing. The normalized frequency of 18 (100 Hz) was observed in both the pressure signals (VL1 and VL2). This frequency is always present in the signals and is attributed to the one-third harmonics of the three-phase rectifier and second harmonics of the grid frequency. It is believed that the frequency will not be available in the system with the generator off.

Other normalized frequencies of 7.5, and 2.9 were also observed in the spectrograms during load acceptances. Assuming a sound wave velocity of 932 m/s, normalized frequencies of 7.5 and 2.9 were calculated for the upstream and downstream pipe lengths to the turbine. The corresponding lengths are 14.5 and 5.6 m, starting from the pressure tank water level to the runner inlet and the runner outlet of the downstream tank water level, respectively. The standing wave frequencies (7.5 and 2.9) start to appear in the system after the transient. During the load acceptance from ML to PL, the normalized frequency of 7.5 was observed to appear first at the location VL1 and then VL2. Significant PSD strength for another standing wave frequency of 2.9 was observed only at location VL1. During the load acceptance from ML to BEP, the normalized frequency of 7.5 and 2.9 was observed to appear simultaneously at the locations VL1 and VL2. Again the normalized frequency of 7.5 was observed to appear first at the location VL1 and then VL2 during load acceptance from ML to HL. The maximum PSD strength of the standing wave frequencies of 7.5 and 2.9 was observed at location VL1.

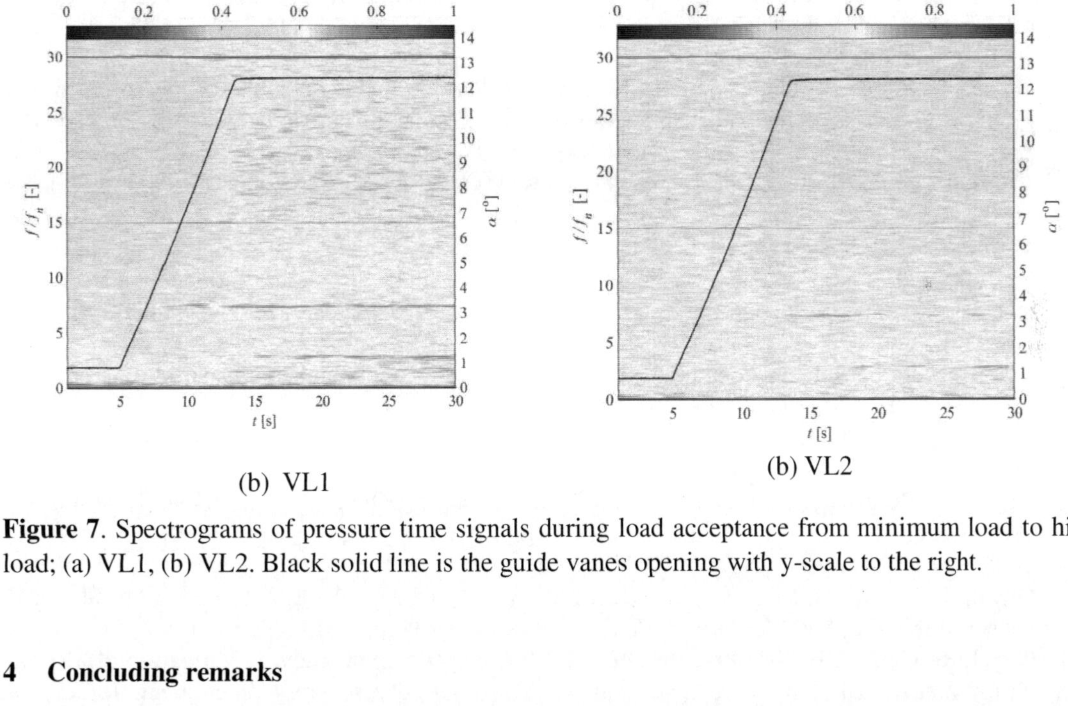

(b) VL1 (b) VL2

Figure 7. Spectrograms of pressure time signals during load acceptance from minimum load to high load; (a) VL1, (b) VL2. Black solid line is the guide vanes opening with y-scale to the right.

4 Concluding remarks

The measurements for load acceptances operations of Francis turbine from the minimum load were performed to investigate the pressure variation and dominant frequencies of fluctuation in the vaneless space. The oscillations in generator torque were observed due to fluctuations in runner rotational speed during load acceptance from ML to BEP. Head was observed as the most affected parameter during load acceptances and the range of head variation was 11.76 to 12.64 m. The maximum pressure fluctuations (± 2.1%) in the vaneless space were observed for the load acceptance condition from ML to PL. The absolute pressure in the vaneless space was ranging from 148 to 178 kPa for all cases. Spectrograms of the pressure signals showed that the PSD strength of blade passing frequency was

IOP Conf. Series: Journal of Physics: Conf. Series **1042** (2018) 012009 doi:10.1088/1742-6596/1042/1/012009

most dominant in the turbine vaneless space. The maximum strength was observed during the load acceptance from ML to HL due the reduced space vaneless between the guide vanes trailing edges and runner blades leading edges. The normalized frequencies of 7.5 and 2.9 were observed and explained as resulting from standing waves in the system due to the downstream and upstream boundary conditions. The maximum PSD strength of the signals was observed at the location VL1 which is located near the beginning of the spiral casing.

Acknowledgement

Rahul Goyal and Prof. Michel Cervantes are grateful to the Swedish water power center (SVC) for the financial support. The authors would like to thank the Norwegian Hydropower Centre (NVKS) for the financial support. The measurements have been carried out in collaboration between LTU, Sweden (NTNU, Norway) and IIT Roorkee. The author's gratitude also goes to the staff of water power laboratory, NTNU, Norway.

References

[1] Gjengedal, T., Integration of Wind Power and the Impact on Power System Operation, *IEEE Large engineering Systems Conference on power Engineering*, May 7-9, 2003, pp. 76-83. ISBN: 0-7803-7863-6; doi:10.1109/LESCPE.2003.1204683.

[2] Goyal, R., Gandhi, B. K., 2017, Review of hydrodynamics instabilities in Francis turbine during off design and transient operations, *Renewable Energy*, **116**, pp. 697-709. doi: 10.1016/j.renene.2017.10.012.

[3] Goyal, R., Gandhi, B. K., Cervantes, M. J., 2017, Particle Image velocimetry measurements in Francis turbine: A review and application to transient operations, *Renewable and Sustainable Energy*, **81 (2)**, pp. 2976-2991, doi:10.1016/j.rser.2017.06.108.

[4] Amiri, K., Mulu, B., Raisee, M., and Cervantes, M. J., 2015, Unsteady Pressure Measurements on the Runner of a Kaplan Turbine during Load Acceptance and load Rejection, *Journal of Hydraulic Research*, **54 (1)**, pp. 56-73.

[5] IEC 60041, 1991-1, Field acceptance tests to determine the hydraulic performance of hydraulic turbines, storage pumps, and pump-turbines, *International Electrotechnical Commission*, Third Edition.

[6] IEC 60193, 1999-11, Hydraulic turbines, storage pumps and pump-turbines – Model acceptance tests, *International Electrotechnical Commission.*

[7] ASME PTC 18, 2011, Hydraulic Turbines and Pump-Turbines, Performance Test Codes, Three Park Avenue, New York, NY 10016 USA.

[8] Goyal, R., Bergan, C., Cervantes, M. J., Gandhi B. K., Dahlhaug. O. G., 2016a, Experimental investigation on a high head model Francis turbine during load rejection, *IOP Conf. Series: Earth and Environmental Science* **49**:082004.

[9] Goyal, R., Cervantes, M. J., Gandhi, B. K., 2016, "Vortex rope formation in a high head model Francis turbine", *ASME Journal of Fluids Engineering*, 139, pp.041102-1-14.

[10] Goyal, R., Gandhi, B. K., Cervantes, M. J., 2017, Experimental study of Mitigation of a spiral vortex breakdown at high Reynolds number under an adverse pressure gradient, *Physics of Fluids* (AIP), **29**, pp. 104104-1-19. oi:10.1063/ 1.4999123

IOP Conf. Series: Journal of Physics: Conf. Series **1042** (2018) 012010 doi:10.1088/1742-6596/1042/1/012010

Impact of Modelling Details on the Generation Function for a Norwegian Hydropower Producer

M N Hjelmeland[1], A Helseth[2] and M Korpås[1]

[1] Department of Electric Power Engineering, NTNU Norwegian University of Science and Technology, Trondheim, Norway.
[2] SINTEF Energy Research AS, Trondheim, Norway.

Abstract. In the following work the generation function relating water discharge to power output for a hydropower station is considered. The conventional approach to modelling the generation function has been to simplify it as a concave function. In the future it is expected that the stations will operate at more varied power outputs, motivating the need for a more detailed modelling of the generation function. Therefore, an investigation on how the nonconcave generation function may be simplified, without having to lose the function's inherent geometric shape, is performed. A greedy algorithm and a Minimum Least Square Error (MLSE) approach is used. It was found that the function can be reduced until a point where the changes become too prominent. A benchmark test against the conventional modelling approach found that it is important with a detailed modelling of the generation function when environmental constraints, such as minimum discharge, are included.

1. Introduction

The following work is founded on recent advancements in hydropower scheduling methods that enables a higher level of modelling details. There has been a fundamental change in the electricity sector with large shares of renewables entering the market and increased cross-border exchange. It is expected that this will lead to increased demand of flexible power plants that can provide ancillary services [1]. Given a market environment where these services can be traded across the European countries it is estimated that there is a income potential for Norwegian hydropower producers [2]. Due to the flexibility storable hydropower possesses they have an opportunity to provide many of the ancillary services at a competitive cost. However, in order for them to adapt to this changing market environment and provide these services it is evident that adequate decision support tools are needed to optimally allocate generation between different markets. Earlier studies have shown that conventional scheduling algorithms based on linear programming (LP) tend to overestimate the amount of capacity reserves that the hydropower producer can deliver [3]. There is also a tendency to increase the environmental restrictions on new concessions for hydropower systems in Norway, leading to more complex constraints in the optimization problem. Developing more advance methods should there be of priority to include these constraints and capture the added income potential by participating in multiple-markets.

2. Hydropower Scheduling

Due to the uncertainty of inflow, energy prices and the possibility of storing water over longer time periods, the hydropower scheduling problem is generally handled as multistage stochastic programming problem. An important element of hydropower scheduling is the water value; even though inflow to the reservoirs are free, there is an opportunity cost associated with the water stored in the reservoirs. The producer can choose to use the water for selling energy today and earn an immediate income or store it for future use with a potential higher profit. In terms of hydropower scheduling, this opportunity cost is referred to as the water value.

Given the complex nature of the hydropower scheduling problem, it can be divided into three problems with different time horizon and modelling details [4]. The Long-Term Hydropower Scheduling (LTHS) problem aims at solving a fundamental market problem that provides energy price forecasts and aggregated water values [5] [6]. The Medium-Term Hydropower Scheduling (MTHS) problem can then use the aggregated water values as end statement and price forecasts to provide individual water values for each reservoir in a local hydropower system [7]. Further, these individual water values can be used as an end statement for the Short-Term Hydropower Scheduling (STHS) problem that has high modelling details, e.g. modelling of individual generators and water courses. The time resolution is fine, typically hourly for a period of one or two weeks. For liberalized electricity markets this problem aims at first generating bids to the power market and after the market clearing perform a re-optimization that incorporates any commitments that occurred in the clearing [8].

2.1. Multistage Stochastic Optimization

This paper depicts an application that will be used in further studies on the MTHS problem. A time horizon of 1-2 years is typical for this problem type, using weekly decision stages as the inflow is highly correlated on a weekly basis. The problem is cast as a Stochastic Dynamic Programming (SDP) problem, with weekly decision stages. To circumvent the "curse of dimensionality" associated with conventional (S)DP problems, a method called Stochastic Dual Dynamic Programming (SDDP) is widely used [9]. The method approximates the expected profit function (EPF), also referred to as cost-to-go function for minimization problems. One of the drawbacks with the method is that it requires the decision stages to be modelled as LP problems, ruling out the possibility of modelling nonlinearities without simplifications. A recent extension of the method has been proposed called the Stochastic Dual Dynamic integer Programming (SDDiP) method [10]. The method can solve nonlinear problems with finite convergence. However, it comes at a cost of added computation time and some adjustments on the structure of the problem, i.e. it requires all the state variables to be binary. Previous work has tested the method on a MTHS problem with promising results, compared to the SDDP method [11]. With this in mind, the following work investigates how the generation function of a hydropower plant can be modelled in the SDDiP framework and how it can be simplified to reduce the computational burden.

2.2. Contributions

The following paper applies a greedy algorithm and a Minimum Least Square Error (MLSE) approach to simplify the generation function of a hydropower station. The greedy algorithm is based on the idea of reducing the number of points on a bid curve in [12]. The approach is tested on two real power station equivalents and the importance of detailed modelling of the generation function is shown. The following section outlines the hydropower scheduling problem, followed by case studies and results.

3. Medium-Term Hydropower Scheduling

The problem aims at allocating resources simultaneously in an energy market and a capacity market. For the sake of simplification, only uncertainty of inflow is considered. The weekly decision problem can be written on dense form as

$$\max_{(x_t, y_t)} \quad f_t(x_t, y_t) + \alpha_t(x_t) \tag{1}$$

$$\text{s.t.} \quad Wx_t + Gy_t = h_t(\xi_t) - Hx_{t-1} \tag{2}$$

$$By_t = 0 \tag{3}$$

$$Cy_t - Dx_t \geq 0 \tag{4}$$

$$Cy_t + Dx_t \leq Cy^{\max} \tag{5}$$

$$x_t, y_t \in Y_t \tag{6}$$

$$x_t \in \mathbb{R}^{k_1} \cdot \mathbb{Z}^{k_2}, y_t \in \mathbb{R}^{l_1} \cdot \mathbb{Z}^{l_2}, \tag{7}$$

where the objective function (1) consists of a present profit function, $f_t(x_t, y_t)$, and the expected future profit function, $\alpha_t(x_t)$. x_t and y_t are respectively the state and stage variables. State variables carry information between stages, e.g. reservoir levels, whereas the stage variables only represent variables within the stages. The inflow is given by the function $h_t(\xi_t)$, where ξ_t is the normalized inflow. The matrices W, G, H, B, C, D are of suitable dimensions, representing the given hydropower system. The time-linking constraint (2) includes all reservoir balances and generator state time-couplings. The energy balance is given by (3) and the constraints describing the system's ability to provide capacity reserves is given in (4) and (5). The generation function and other miscellaneous constraints, such as bypass limits, spillage limits and other system dependent constraints are included in (6).

3.1. Generation function

The generation function describes how the hydropower station's discharge relates to the output power, usually modelled as a piecewise linear function. The function is typically computed when the hydropower station is built. This function provides data input to the scheduling models and as a control that the station meets the efficiency as promised by the contractors. As the efficiency may deteriorate over time, the more recent the measurements, the better. Advantageously, a system with continuous measurement could give the best result in terms of describing the generation function most accurately. The methods proposed in this work will work irrespective of the measurement approach, it would just require some prepossessing of the data.

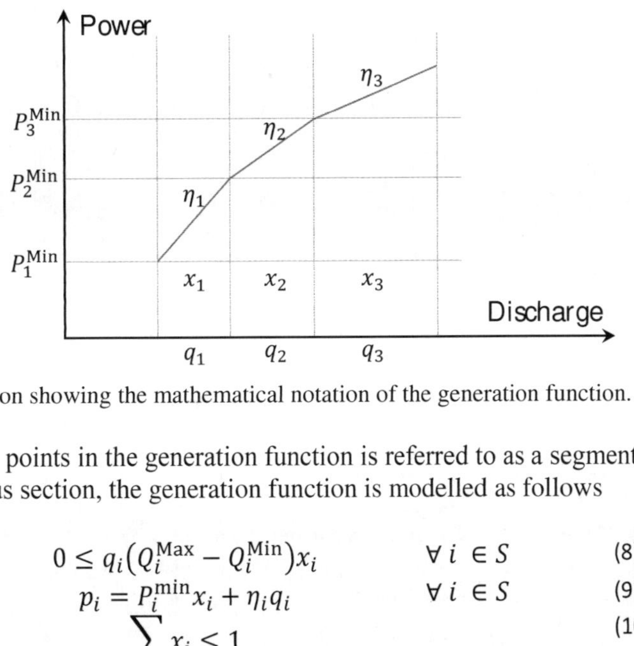

Figure 1: Illustration showing the mathematical notation of the generation function.

The line between two points in the generation function is referred to as a segment. Extending on the notation from previous section, the generation function is modelled as follows

$$0 \leq q_i(Q_i^{Max} - Q_i^{Min})x_i \qquad \forall i \in S \qquad (8)$$

$$p_i = P_i^{min}x_i + \eta_i q_i \qquad \forall i \in S \qquad (9)$$

$$\sum_{i \in S} x_i \leq 1 \qquad (10)$$

$$p = \textstyle\sum_{i \in S} p_i, \qquad (11)$$

where q_i is the discharge (in Mm^3) for segment i, Q_i^{Min} and Q_i^{Max} are the minimum and maximum discharge for each segment, p_i is the generation (in MW) for each segment, P_i^{Min} is the minimum generation for each segment, x_i is a binary variable indicating whether the segment i is active or not, p and p_i is, respectively, the overall generation and generation for the individual segments in the set S of segments. An illustration of the generation function is given in Figure 1. Observe that even though the illustration depicts a concave function the method also applies to the nonconcave case, which is an important aspect in this paper. A concave generation function would imply that the first segment has the best efficiency, i.e. best power output for a given discharge, while the other segments are decreasingly

IOP Conf. Series: Journal of Physics: Conf. Series **1042** (2018) 012010 doi:10.1088/1742-6596/1042/1/012010

worse. This case would need no binary variables to ensure that the efficiency of each segment is used correctly, since the model will always use the segment with the best efficiency first and the rest followingly.

3.1.1. Area removal algorithm

The following section describes the greedy algorithm used to reduce the number of segments in the generation function. The idea is taken from [12], where the authors apply the method on bid curves for a hydropower producer. The underlying idea is that you should remove points in the function that alter its geometric shape as little as possible. Illustration given in Figure 2.

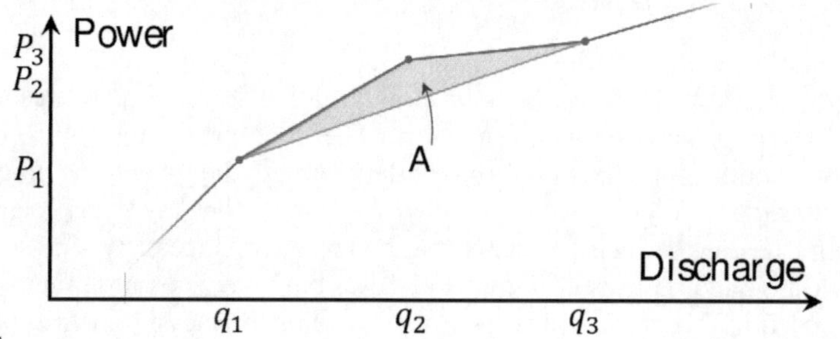

das

Figure 2: The figure shows a part of the generation function and how the area (A) is computed between three adjacent points.

The algorithm computes the area between two adjacent point for all the points on the curve. The point that has the lowest computed area is removed. This procedure is repeated until a desired amount of points are removed.

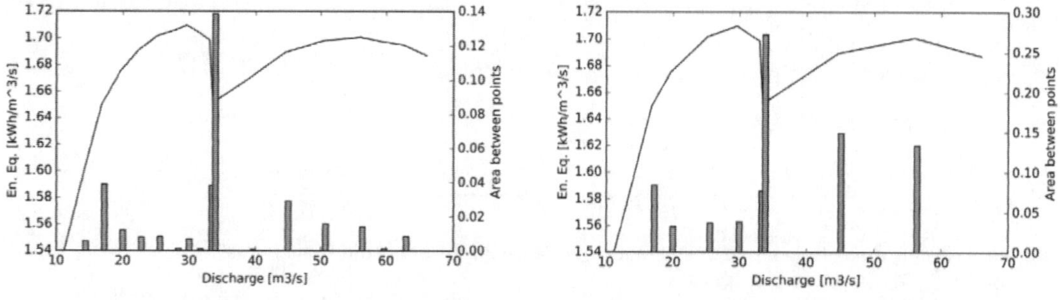

Figure 3: Illustration of the generation function. Right axis indicates the area of the triangle computed with the two adjacent points. Note that we are illustrating the energy equivalent [$\mathbf{kWh/m^3}$]. This is done to get a better visualization of the curve. Left: The original generation function. Right: The generation function after eight points have been removed by the area removal technique.

Figure 3 shows the generation function and the area between the adjacent points for the original function and for the function when eight points have been removed. As expected the geometric shape of the function is very much kept, even though half of the points on the curve has been removed. Below, in Figure 4, the accumulated area that is removed from the function is plotted. It is evident that there are many points on the curve that can be removed with almost no impact. After a certain threshold the accumulated area increases significantly, giving an initial indication for how much the function can be reduced.

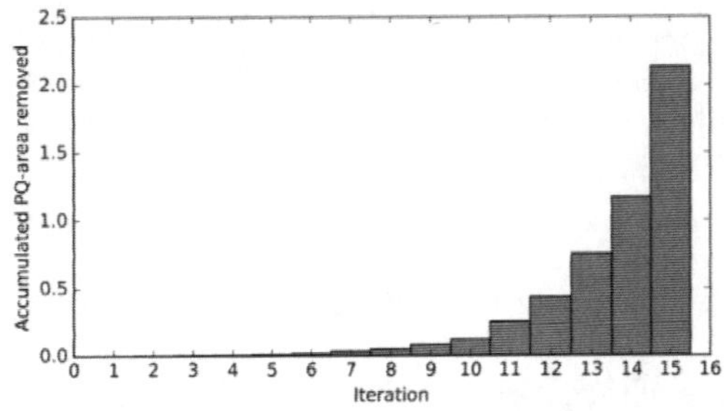

Figure 4: Bar graph of the accumulated area between adjacent points in the generation function that are removed. As illustrated, there are many points that can be removed with little impact on the shape of the curve, followingly the more points removed, the more significant this impact becomes.

3.1.2. Segmented regression method

Segmented regression is applied to reduce the number of points in the generation function. This is done with the R library *segmented* [13]. It is an iterative approach where the user suggests starting breakpoints. Being a statistical approach, it would be beneficial with a lot of data to fit the curve. In such sense, the approach would be well suited for a system that does continuous measurements. Nonetheless, the approach provides results well fitted to the original data, as seen in Figure 5.

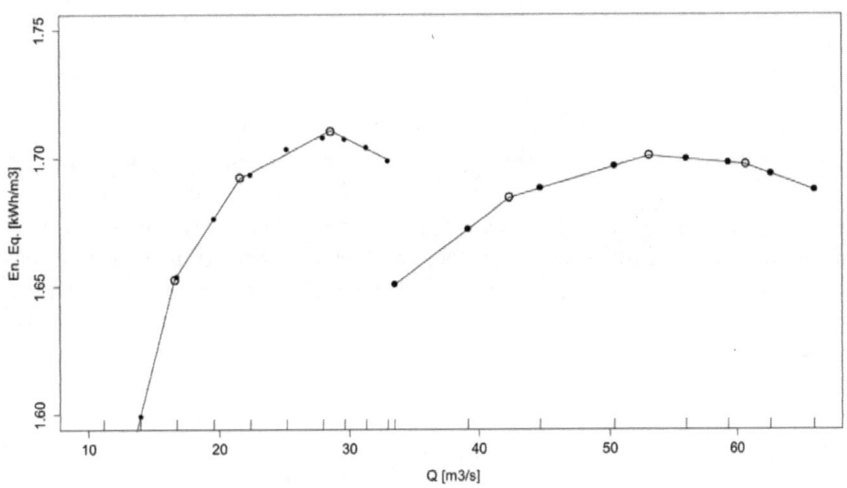

Figure 5: The generation function is given by the filled dots. The line going through the circles are the results from the segmented regression. Observe that there are two generators in the power station, each fitted individually.

4. Case Study

Two power stations are investigated in the study. The first, Station A, consists of a single generator unit with 45 MW installed capacity. The second, Station B, consists of two generator units, each with 210 MW installed capacity. Both stations only have Francis turbines installed.

The two systems are solved with a time horizon of two years with weekly uncertainty of inflow. For simplicity, the energy prices and capacity reserve prices are modelled as deterministic. The week is then separated into 21 time periods, three for each day indicating morning, mid-day and evening.

The SDDiP approach is applied, described in [10] [11], to solve the hydropower scheduling problem. The problem is solved with the original generation function and an EFP function is obtained, which in turn is used to simulate with the different reductions of the generation function. For each simulation 100 different scenarios are generated that are used to evaluate the expected results, for all simulations. To compare the results to what is normal approach for this problem type in operative models today, a case

IOP Conf. Series: Journal of Physics: Conf. Series **1042** (2018) 012010 doi:10.1088/1742-6596/1042/1/012010

where the generation function is concave is performed. Lastly, some tests are performed on how Station A would be operated given that there is a time period in the summer months that requires a minimum discharge, due to environmental constraints.

5. Results and Discussion

The following section outlines some results from the case studies and discusses its implications.

Table 1 and Table 2 shows some of the results obtained from the different simulations. The area removal algorithm (A. Rem.), segmented regression method (Seg. Reg.) and Concave case (Conc.) was run. The tables depict the computation time of each simulation, the expected objective value, deviation in shape of the duration curve and total generation from the original problem, given by the first column. The second row of the table indicates how many points on the curve has been removed from an original total amount of 18 and 11 in System A and B, respectively.

Table 1: Results from System A. Simulation time is given in seconds, whereas the other values are given as a percentage. Val. represents the expected objective value, Dur. indicates the percentage deviation of the duration curve compared to the original problem. Gen. is the expected generation for the two years.

	A. Rem						Seg. Reg.				Conc.
	0	2	4	6	8	10	8	9A	9B	10	
Time	920	895	763	668	614	523	555	522	521	475	59
Val.	0.00	0.00	0.01	0.00	0.00	-0.03	-0.01	-0.02	-0.01	-0.01	2.22
Dur.	0.00	0.67	0.69	0.67	1.33	2.62	3.10	4.39	4.77	4.28	13.02
Gen.	0.00	-0.01	0.02	-0.01	-0.30	-0.15	-0.14	0.46	0.24	-0.11	0.34

As expected the computation time drops significantly with the number of points being removed. This is especially evident for the concave case where there are less constraints and binary variables in the problem. Moreover, one can observe that the expected objective is not changed much between the cases. This is due to the fact that points on the generation curve are being removed regardless whether it results in an increase or decrease in overall efficiency. The changes in shape of the duration function increases the more points are removed from the generation function. This is especially clear for the segmented regression case, but not unexpected as seen from Figure 5, where the best operating points shift slightly. Lastly, another expected result occurs where there is a positive change in generation for the concave case.

Table 2: Results from System B. Simulation time is given in seconds, whereas the other values are given as a percentage. Val. represents the expected objective value, Dur. Indicates the percentage deviation of the duration curve compared to the original problem. Gen. is the expected generation for the two years.

	A. Rem.					Seg. Reg.		Conc.
	0	2	4	6	8	6	7	
Time	335	284	238	218	180	197	178	77
Val.	0.00	-0.03	-0.01	-0.05	-0.58	0.01	0.09	2.05
Dur.	0.00	0.16	0.50	0.52	9.99	1.04	2.95	15.09
Gen.	0.00	0.00	-0.04	0.06	-0.76	0.09	0.01	1.02

The duration curve for selected simulations are shown in Figure 6 illustrates some duration curves for System A. The degree of regulation, i.e. the system's ability to store water for usage at a later stage, is very high for the system. Thus, the model will shift as much of the generation as possible to the hours in the winter where the prices are highest. Also, considering that the efficiency of the station is low at low power outputs the model will tend to avoid such operation. The best efficiency of the station is obtained around 180 MW, where a large portion of the generation comes from. Note that the concave generation function is not able to represent the low efficiency at low power outputs, some generation is therefore observed around 60 MW for this simulation. The yellow line shows the duration curve for the simulations with the segmented regression method applied. As expected the generation points are somewhat shifted from the original problem. This illustrates the importance of the generation function modelling on the operation of the power station. This method would, as above-mentioned, get a much more robust representation of the generation function had there been more measurements to fit.

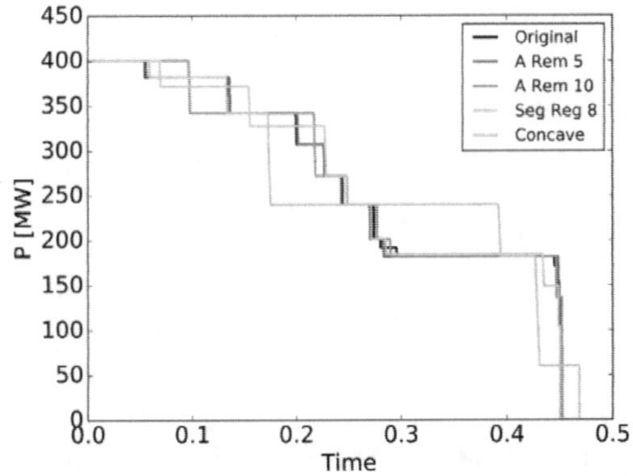

Figure 6: Duration curve for a selection of the results from System A. A. Rem. refers to the Area removal method and how many points are remove. Seg Reg 8 is for the segmented regression and the grey line is for the concave generation function.

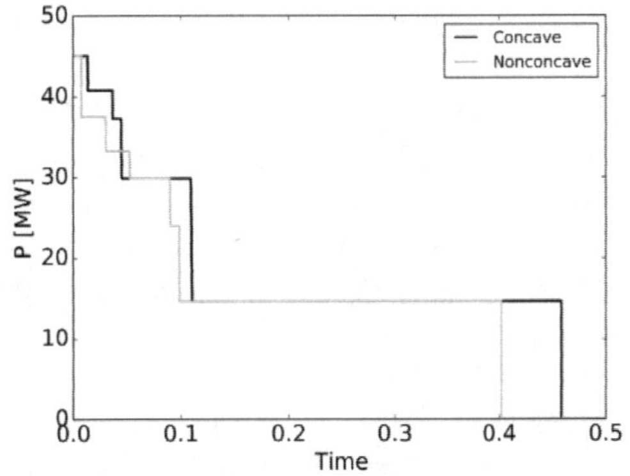

Figure 7: Duration curve for a selection of the results from System B. Minimum discharge limit during summer months.

Figure 7 shows the duration curve when Station B ran with a minimum generation requirement during the summer months. This would result in much less flexibility in terms of storing the water for the hours of the year with highest prices. The simulations were performed as a test to investigate how the system reacts to generation on lower power outputs. The minimum power output for the station is 15 MW, as seen in the figure where most of the power station operation is performed. The tests were performed with the original generation function and the concave one. It can be seen that the shape of the duration curves is very much similar, but with much less generation for the original, nonconcave, problem. This shows a significant drawback with the concave generation function, it overestimates the efficiency of the station and therefore assume that it can generate more electricity than what is possible. For the given case the simulation with concave generation function had an expected generation of 14.1% more than the original one, indicating how much the generation can be overestimated for simplifications of the generation function. For the other simulations this number was lower (1.02% and 0.34%), showing that how the station is operated has an impact on how much detail should be included in the generation function. Since it is expected that the market for ancillary services will grow and increased environmental constraints will occur, more operation of power stations at low power output can be more common in the future.

6. Conclusion and Further Work

In the given paper a segmented regression approach and a greedy algorithm have been proposed and tested to simplify the piecewise linear generation function used to model hydropower stations. Impacts from environmental constraints and how they alter the operation has been examined, demonstrating the importance of a more detailed modelling of the generation function.

The assumption that the generation function is only dependent on discharge is reasonable for stations with high head and less regulation of the reservoirs. On the contrary, if this is not the case further work will investigate methods that can include the head dimension in the generation function.

Acknowledgments
This work was supported by the Research Council of Norway and industry partners under Grant 228731/E20.

Article I. References

[1] P. Cramton, A. Ockenfels, S. Stoft and others, "Capacity market fundamentals," *Economics of Energy & Environmental Policy,* vol. 2, no. 2, pp. 27-46, 2013.

[2] M. Fodstad, M. M. Aarlott and K. T. Midthun, "Value-Creation Potential from Multi-Market Trading for a Hydropower Producer," *Energies,* vol. 11, no. 1, 2018.

[3] M. N. Hjelmeland, M. Korpås and A. Helseth, "Combined SDDP and simulator model for hydropower scheduling with sales of capacity," in *2016 13th International Conference on the European Energy Market (EEM)*, 2016.

[4] O. B. Fosso, A. Gjelsvik, A. Haugstad, B. Mo and I. Wangensteen, "Generation scheduling in a deregulated system. The Norwegian case," *IEEE Transactions on Power Systems,* vol. 14, no. 1, pp. 75-81, 1999.

[5] A. Gjelsvik, B. Mo and A. Haugstad, "Long- and Medium-term Operations Planning and Stochastic Modelling in Hydro-dominated Power Systems Based on Stochastic Dual Dynamic Programming," in *Handbook of Power Systems I*, Eds., Springer Berlin Heidelberg, 2010, pp. 33-55.

[6] O. Wolfgang, A. Haugstad, B. Mo, A. Gjelsvik, I. Wangensteen and G. Doorman, "Hydro reservoir handling in Norway before and after deregulation," *Energy* , vol. 34, no. 10, pp. 1642-1651, 2009.

[7] A. Helseth, M. Fodstad and B. Mo, "Optimal Medium-Term Hydropower Scheduling Considering Energy and Reserve Capacity Markets," *IEEE Transactions on Sustainable Energy,* vol. PP, no. 99, pp. 1-9, 2016.

[8] E. K. Aasgård, G. S. Andersen, S. E. Fleten and D. Haugstvedt, "Evaluating a Stochastic-Programming-Based Bidding Model for a Multireservoir System," 2014.

[9] M.V.F. Pereira and L. M. V. G. Pinto, "Multi-stage stochastic optimization applied to energy planning", Mathematical Programming, vol 52, no. 1-3, pp. 359-375, 1991.

[10] J. Zou, A. Sun and S. Ahmed , "Stochastic dual dynamic integer programming," *Mathematical Programming,* Mar 2018.

[11] M. N. Hjelmeland, J. Zou, A. Helseth and S. Ahmed, "Nonconvex Medium-Term Hydropower Scheduling by Stochastic Dual Dynamic Integer Programming," *IEEE Transactions on Sustainable Energy*, 2018.

[12] E. K. Aasgård, C. Ø. Naversen, M. Fodstad and H. I. Skjelbred, "Optimizing day-ahead bid curves in hydropower production," *Energy Systems,* Feb 2017.

[13] V. M. Muggeo, "Segmented: an R package to fit regression models with broken-line relationships," *R news,* vol. 8, no. 1, pp. 20-25, 2008.